T0309393

NEURONS

A Mathematical Ignition

Series on Number Theory and Its Applications ISSN 1793-3161

Published

Vol. 1 Arithmetic Geometry and Number Theory
edited by Lin Weng & Iku Nakamura

Vol. 2 Number Theory: Sailing on the Sea of Number Theory
edited by S. Kanemitsu & J.-Y. Liu

Vol. 4 Problems and Solutions in Real Analysis
by Masayoshi Hata

Vol. 5 Algebraic Geometry and Its Applications
edited by J. Chaumine, J. Hirschfeld & R. Rolland

Vol. 6 Number Theory: Dreaming in Dreams
edited by T. Aoki, S. Kanemitsu & J.-Y. Liu

Vol. 7 Geometry and Analysis of Automorphic Forms of Several Variables
Proceedings of the International Symposium in Honor of Takayuki Oda on the Occasion of His 60th Birthday
edited by Yoshinori Hamahata, Takashi Ichikawa, Atsushi Murase & Takashi Sugano

Vol. 8 Number Theory: Arithmetic in Shangri-La
Proceedings of the 6th China–Japan Seminar
edited by S. Kanemitsu, H.-Z. Li & J.-Y. Liu

Vol. 9 Neurons: A Mathematical Ignition
by Masayoshi Hata

Series on Number Theory and Its Applications Vol. 9

NEURONS
A Mathematical Ignition

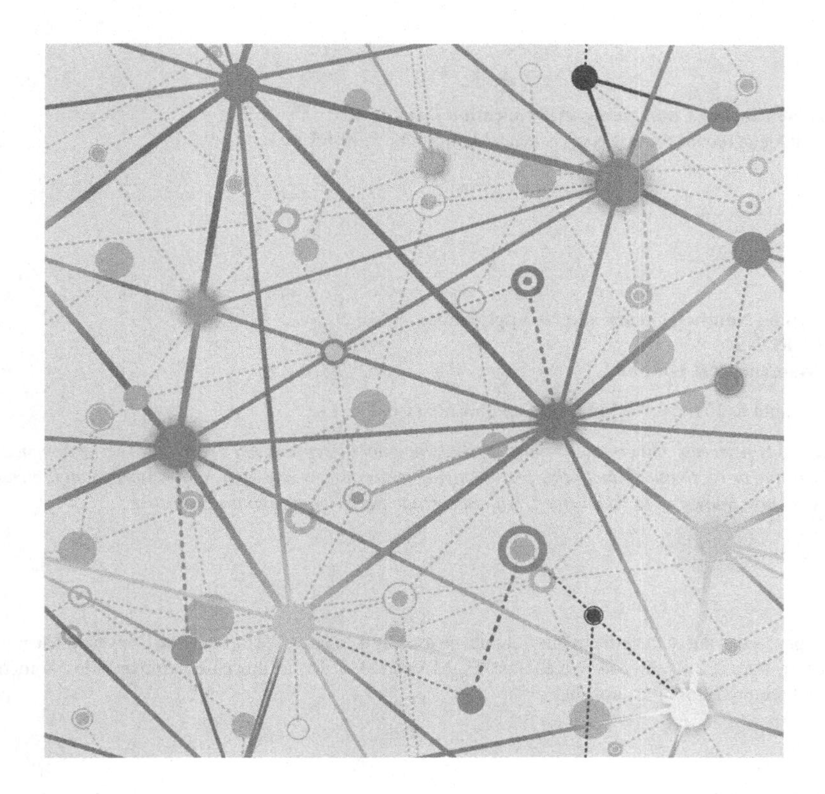

Masayoshi Hata
Kyoto University, Japan

NEW JERSEY • LONDON • SINGAPORE • BEIJING • SHANGHAI • HONG KONG • TAIPEI • CHENNAI

Published by

World Scientific Publishing Co. Pte. Ltd.
5 Toh Tuck Link, Singapore 596224
USA office: 27 Warren Street, Suite 401-402, Hackensack, NJ 07601
UK office: 57 Shelton Street, Covent Garden, London WC2H 9HE

British Library Cataloguing-in-Publication Data
A catalogue record for this book is available from the British Library.

Series on Number Theory and Its Applications — Vol. 9
NEURONS
A Mathematical Ignition

ISBN 978-981-4618-61-8

Printed in Singapore

Preface

The present book is a considerably revised version of materials given by the author in some intensive courses at Hokkaido University, the University of Tokyo and Osaka City University in 1996, and at Yamaguchi University in 2010.

It is said that our brain consists of more than a hundred billion neurons, which are connected mutually by synapses and behave tremendously in its complexity, and that, if one connects all the axons and dendrites in a brain end to end, then it attains a length in excess of a million kilometers. In 1961 E. R. Caianiello published a monumental paper entitled "Outline of a theory of thought-processes and thinking machines" in *Journal of Theoretical Biology*, in which he had strengthened the conviction that the human brain yet obeys dynamical laws that are not necessarily complicated, if one looks at the operation of individual neurons.

This book deals with two mathematical sides of Caianiello's neuronic equations; one is from the point of view of dynamical systems and the other from the number-theoretic point of view. In particular, the latter shows that Caianiello's equations are closely related to some topics in elementary number theory and even in transcendental number theory.

The author has come full circle back to Caianiello's neuronic equations, which were the first research theme in his career. The author expresses deep gratitude to the late Professor Masaya Yamaguti, his supervisor at Kyoto University, who brought him to this subject of research and handed his first paper to the late Professor Caianiello directly at the University of Salerno in the late 1980s.

Each chapter except for Chapter 2 ends with some instructive and mostly original exercises at various levels. Some contain examples, additional results and related topics, which would provide a sense of perspective, and several are used in the text. For almost all exercises the detailed solutions are given in Hints and Solutions at the end of the book. The table of the associated polynomials will help for readers to verify and investigate their relations.

All the figures in the book were drawn by using "Grapher 2.1" bundled with Mac OS X.

We refer to the website of I. I. A.S.S. (`http://www.iiassvietri.it`) for the profile and historical notes about Professor Caianiello.

The author cordially appreciates Professor Shigeru Kanemitsu who provided uninterrupted encouragement and the exquisite title of this book with a twofold significance. The author is also deeply indebted to Professor Michel Waldschmidt for his careful reading of the manuscript and for giving detailed comments and suggestions. Also thanks are due to the staff of World Scientific Publishing Co. for the excellent help and cooperation.

This work is supported by the Japan Society for the Promotion of Science (JSPS) through the "Funding Program for World-Leading Innovative R&D on Science and Technology (FIRST Program)".

Kyoto, JAPAN M. Hata

Contents

Symbols and Notations

In the following definitions arabic numerals indicate the page numbers where they appear in the book for the first time.

$\mathrm{Orb}_f(x)$	orbit of x by the mapping f, 1
$\mathbb{N}$	set of positive integers, 1
$\mathrm{Fix}(f)$	set of fixed points of the mapping f, 1
$\mathrm{Per}_m(f)$	set of periodic points of the mapping f with period m, 1
$\mathrm{Per}(f)$	set of fixed and periodic points of the mapping f, 1
$\mathbb{R}$	set of real numbers, 3
$\mathbb{Z}$	set of integers, 4
$[x]$	integral part of x, 4
$\{x\}$	fractional part of x, 4
$\#E$	number of elements in the set E, 5
$m \equiv n \pmod p$	m and n are congruent modulo p, 5
$\gcd(p, q)$	greatest common divisor of p and q, 5
$\mathfrak{M}$	set of piecewise monotonically increasing functions, 6
$\tilde{f}(x)$	left limit function of f, 6
$\{0, 1\}^{\mathbb{N}}$	set of sequences consisting of 0 and 1, 8
$C^3(I)$	set of three times continuously differentiable functions on I, 9
$S(f)(x)$	Schwarzian derivative of the function f, 9
$\mathrm{H}(x)$	Heaviside step function, 13
$\psi_{a,b,c}(x)$	piecewise linear function with three parameters a, b and c, 15
$\lVert \boldsymbol{x} \rVert_\infty$	maximum norm of the vector $\boldsymbol{x}$, 16
$\mathfrak{B}$	set of some discontinuous functions defined on $[0, 1]$, 19
$f^*(x)$	reflection of the function f, 19
c_{-n}	inverse image of the discontinuity point c by f^n, 21
$\epsilon(x)$	address of x, 21
$\eta(x)$	itinerary of x, 21

$\rho(f)$	rotation number of the function f, 23
$\Theta_f(x,t)$	generating function of the itinerary of x, 28
$\mathbb{Z}[[t]]$	ring of formal power series in t with integral coefficients, 28
$\overline{E}$	closure of the set E, 33
$f(E)$	image of the set E under the mapping f, 33
$\mathrm{Int}\, E$	interior of the set E, 33
$a_n(f)$	number of subintervals constituting the closure of $f^n(J)$, 33
$X \setminus Y$	set of elements of X not belonging to Y, 34
$X \prec Y$	$X \cap Y = \emptyset$ and $\sup X \le \inf Y$, 37
$\mathfrak{E}(p/q)$	subset of $\mathfrak{B}_q$, 38
$\varphi(q)$	Euler's totient function, 38
$Q_{p,q}(t)$	basic pattern polynomial of $\mathfrak{E}(p/q)$, 39
$\mathfrak{E}^{\phi}(p/q)$	subset of $\mathfrak{E}(p/q)$, 40
$\mathfrak{E}^{*}(p/q)$	subset of $\mathfrak{E}(p/q)$, 40
$\mathfrak{E}^{0}(p/q)$	subset of $\mathfrak{E}(p/q)$, 40
$\lvert E \rvert$	one-dimensional Lebesgue measure of $E \subset \mathbb{R}$, 40
$\mathcal{F}_n$	Farey series of order n, 45
$\mathbb{Q}$	set of rational numbers, 46
$\mathcal{I}_n$	set of Farey intervals of order n, 47
$\lfloor x \rfloor$	floor function; the largest integer not greater than x, 49
$\lceil x \rceil$	ceiling function; the smallest integer not less than x, 49
$\mathfrak{S}(p/q)$	itinerary of $f(0)$ for $f \in \mathfrak{E}(p/q)$, 51
$\mathfrak{S}^{*}(p/q)$	tail inversion sequence of $\mathfrak{S}(p/q)$, 51
p_n/q_n	lower approximation sequence, 53
r_n/s_n	upper approximation sequence, 53
$\ell_n(\kappa)$	positive integer defined in (iii) in Section 5.6, 53
$\mathbb{N}^{\mathbb{N}}$	set of sequences consisting of positive integers, 54
$\vartheta(\kappa)$	limit superior of the sequence s_n/q_n, 55
ϖ	constant defined as $(3-\sqrt{5})/2$, 55
$H_\mu(z)$	power series studied by Hecke, 71
$C[0,1]$	set of continuous functions on $[0,1]$, 72
Ω_f	limit set for $f \in \mathfrak{B}_\infty$, 75
κ_n	maximal length of subintervals constituting the image $f^n(J)$, 76
e_n	infimum of the difference between x and $f^n(x)$, 77
$\omega_f(x)$	ω-limit set of x for $f \in \mathfrak{B}_\infty$, 79
Λ	set of parameters (a,b,c) satisfying $\psi_{a,b,c} \in \mathfrak{B}(c)$, 85
$P_{p,q}(w,z)$	associated polynomial in w and z, 86
$P_{p,q}^{+}(w,z)$	another associated polynomial in w and z, 86

$M_\mu(w,z)$	Hecke-Mahler series, 92
$Q_{1/\mu}(w,z)$	infinite series associated with the Hecke-Mahler series, 92
$\mathscr{H}^s(X)$	Hausdorff s-dimensional outer measure of $X \subset \mathbb{R}^n$, 96
$\dim_{\mathrm{H}} X$	Hausdorff dimension of $X \subset \mathbb{R}^n$, 96
φ	constant defined as $(\sqrt{5}+1)/2$; the golden ratio, 99
F_n	nth Fibonacci number, 100
$r_{a,b}$	upper bound of c for fixed a and b, 103
$u_n(c)$	orbital function with initial point 0, 103
$v_n(c)$	orbital function with initial point 1, 103
$\kappa_{a,b}$	non-negative number defined as $\log a/\log(a/b)$, 104
A_n	set of discontinuity points of $u_n(c)$, 105
B_n	set of discontinuity points of $v_n(c)$, 105
$U_n(c)$	itinerary function corresponding to $u_n(c)$, 107
$V_n(c)$	itinerary function corresponding to $v_n(c)$, 107
ρ^*	orbital average, 111
$\varDelta(p/q)$	closed interval in c-axis in which the periodic case occurs, 117
$\varGamma_{a,b}$	residual set for given a, b with $a \geq b$ and $0 < b < 1$, 123
$\alpha(p/q)$	right endpoint of the interval $\varDelta(p/q)$, 125
$\beta(p/q)$	left endpoint of the interval $\varDelta(p/q)$, 125
$\mathrm{dist}\left(\frac{p}{q},\frac{r}{s}\right)$	distance between $\varDelta(p/q)$ and $\varDelta(r/s)$, 126
$\varrho\left(\frac{p}{q},\frac{r}{s}\right)$	ratio of $\left\lvert\varDelta\left(\frac{p+r}{q+s}\right)\right\rvert$ to $\mathrm{dist}\left(\frac{p}{q},\frac{r}{s}\right)$, 131
$\Gamma(z)$	gamma function, 136
$\Psi(z)$	digamma function, 136
C	Euler's constant, 136
$R_{a,b}(c)$	rotation number function, 139
$C(x)$	Cantor's function, 141
$\alpha_{p/q}(w,z)$	$\alpha_{p/q}$-leaf function, 143
$\beta_{p/q}(w,z)$	$\beta_{p/q}$-leaf function, 143
$A^+_{r/s}(w,z)$	limit of α-leaf functions from above, 146
$A^-_{p/q}(w,z)$	limit of α-leaf functions from below, 148
$B^+_{r/s}(w,z)$	limit of β-leaf functions from above, 153
$B^-_{p/q}(w,z)$	limit of β-leaf functions from below, 155
$\mathbb{C}[[x]]$	ring of formal power series in x with complex coefficients, 165
$\omega(\gamma)$	irrationality exponent of the irrational number γ, 166
ϑ^*	limit superior of the sequence s_{n+1}/s_n, 169

Chapter 1

Basics of Discrete Dynamical Systems

We introduce some basic concepts and notations in "discrete dynamical systems", as well as a few simple examples related to "elementary number theory". The readers familiar with the theory of dynamical systems may skip this introductory chapter.

1.1 Orbits and Periodic Points

Given a set X, we consider a mapping f from X into itself. Note that f is not necessarily *onto*. Since $f(X) \subset X$, the n-fold composition of f can be defined on X for all $n \geq 1$, which is denoted by

$$f^n(x) = \underbrace{f \circ \cdots \circ f}_{n \text{ times}}(x).$$

The *orbit* of $x \in X$ by f is the subset of X defined by

$$\mathrm{Orb}_f(x) = \{x, f(x), f^2(x), \ldots, f^n(x), \ldots\}, \tag{1.1}$$

which is called more precisely the forward orbit of x. This definition is static but it is natural to regard (1.1) as a sequence in X, in other words, an element in $X^{\mathbb{N}}$ rather than a subset of X. In this dynamic sense f defines a discrete dynamical system on X and the first point x in $\mathrm{Orb}_f(x)$ is called the *initial point* of the orbit.

The point $x \in X$ is called a *fixed point* of f if $f(x) = x$. The set of all fixed points of f is denoted by $\mathrm{Fix}(f)$. The point $x \in X$ is called a *periodic point* with period $m \geq 2$ if $f^m(x) = x$ and $f^k(x) \neq x$ for any $1 \leq k < m$. The set of all periodic points with period m is denoted by $\mathrm{Per}_m(f)$. Moreover it may be convenient to define $\mathrm{Per}_1(f) = \mathrm{Fix}(f)$ and

$$\mathrm{Per}(f) = \bigcup_{n \geq 1} \mathrm{Per}_n(f),$$

which means the set of all fixed and periodic points of f. If $x \notin \mathrm{Per}(f)$ and $f^m(x) \in \mathrm{Per}_n(f)$ for some $m, n \geq 1$, then x is said to be an *eventually* periodic point. In particular, there are no eventually periodic points if f is one-to-one.

Example 1.1. Consider the mapping $f : \mathbb{N} \to \mathbb{N}$ defined by

$$f(n) = \begin{cases} n/2 & (n : \text{even}), \\ 3n+1 & (n : \text{odd}). \end{cases}$$

Clearly the point 1 is periodic with period 3, because $f(1) = 4, f(4) = 2$ and $f(2) = 1$. It is conjectured that, for any $n \in \mathbb{N}$ there exists $m \in \mathbb{N}$ satisfying $f^m(n) = 1$. This is known as "the Collatz problem", because L. Collatz first proposed this problem in 1937. This is also called "the $3n + 1$ problem", "the Ulam conjecture", "Kakutani's problem", "the Thwaites conjecture" or "the Syracuse problem". If the conjecture would be true, then every positive integer except for 1, 2 and 4 becomes an eventually periodic point. Currently this is still open and checked by computers up to $n \leq 2.3 \times 10^{21}$.

By definition $\mathrm{Per}_n(f)$ is a subset of $\mathrm{Fix}(f^n)$. More precisely we have the following

Theorem 1.1. *The set* $\mathrm{Fix}(f^n)$ *is expressed as a disjoint union, as follows:*

$$\mathrm{Fix}(f^n) = \bigcup_{d|n} \mathrm{Per}_d(f) \tag{1.2}$$

where d runs over all divisors of n.

Proof. It suffices to show the inclusion relation

$$\mathrm{Fix}(f^n) \subset \bigcup_{d|n} \mathrm{Per}_d(f).$$

For any $x \in \mathrm{Fix}(f^n)$ let $k \in [1, n]$ be the smallest integer satisfying $f^k(x) = x$. Putting $n = sk + r, 0 \leq r < k$, we have

$$\begin{aligned} x &= f^n(x) \\ &= f^{sk+r}(x) \\ &= f^r \circ \underbrace{f^k \circ \cdots \circ f^k}_{s \text{ times}}(x) \\ &= f^r(x). \end{aligned}$$

Thus $r = 0$ and hence k is a divisor of n. □

The formula (1.2) may be reminiscent of the following analogy with Möbius inversion formula:

$$\mathrm{Per}_n(f) = \bigcup_{d|n} \mu(d)\mathrm{Fix}(f^{n/d}),$$

where $\mu(n)$ is the Möbius function. However we must give attention to the "minus" operation of sets, which is not merely the set difference. For example, in the formula

$$\mathrm{Per}_6(f) = \mathrm{Fix}(f^6) - \mathrm{Fix}(f^3) - \mathrm{Fix}(f^2) + \mathrm{Fix}(f),$$

any fixed point is counted twice in "+" sign and twice in "−" sign; so they are pushed out.

1.2 Orbits in the Sense of Kuratowski

In 1924, K. Kuratowski introduced the equivalence relation "$\sim$" in X as follows: $x \sim y$ if $f^m(x) = f^n(y)$ for some non-negative integers m, n, or equivalently, if

$$\mathrm{Orb}_f(x) \cap \mathrm{Orb}_f(y) \neq \emptyset.$$

He remarked that the corresponding equivalence classes can be used in solving Abel's functional equation

$$\phi(f(x)) = \phi(x) + 1 \tag{1.3}$$

for a given function $f(x)$. This equivalence relation was rediscovered by G. T. Whyburn, who named them "orbits". See Targonski (1981) for the details. To distinguish them from our orbits defined in the previous section, we call them "orbits in the sense of Kuratowski".

For example, the functional equation (1.3) for $f(x) = x^2$ possesses a continuous solution

$$\phi(x) = \frac{\log\left|\log|x|\right|}{\log 2} \qquad (x \neq 0, \pm 1).$$

Note that 0, 1 are fixed points and -1 is an eventually fixed point of the quadratic map $f : \mathbb{R} \to \mathbb{R}$.

The following lemma is one of the fundamental properties of Kuratowski's orbits and will be used in Section 7.5.

Lemma 1.1. *Let f be a one-to-one mapping from an interval into itself. Then there exist uncountable different orbits of f in the sense of Kuratowski.*

Proof. Suppose, on the contrary, that there are only countably many orbits in the sense of Kuratowski. If $x \sim y$, then x can be written as $f^n(y)$ for some $n \in \mathbb{Z}$, because f is one-to-one. This means that there are only countably many initial points belonging to the same equivalence class. This is a contradiction, because the union of countably many countable sets is countable. □

1.3 Piecewise Linear Maps

Piecewise linear and piecewise monotone functions are very important classes in the theory of discrete dynamical systems. A function ψ defined on an interval I is said to be *piecewise linear* provided that I can be divided into finitely many subintervals, as mutually disjoint subsets, so that the restriction of ψ to each subinterval is linear. Similarly we define a *piecewise monotonically increasing* function, which is strictly monotonically increasing and continuous on each subintervals. Usually the number of subintervals is taken to be as small as possible.

We denote the integral and fractional parts† of x by $[x]$ and $\{x\}$ respectively. A typical example of piecewise linear and piecewise monotonically increasing function which maps $[0, 1)$ into itself, is $\psi(x) = \{\alpha + \beta x\}$ when $\beta > 0$. Such a mapping is called a beta transformation if $\beta > 1$.

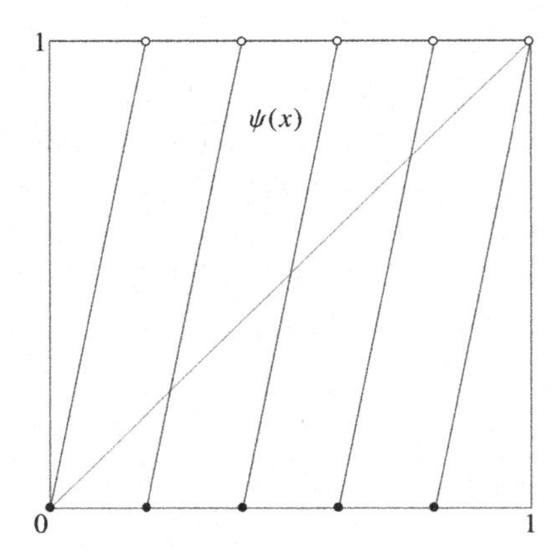

Fig. 1.1 $\psi(x) = \{5x\}$ has four fixed points in $[0, 1)$.

† The fractional part $\{x\} = x - [x]$ is sometimes called the *sawtooth function*. Since curly brackets are also used to delimit sets, we try to append words like "set" or "sequence" to it to avoid any possibility of confusion.

Example 1.2. Let $m \geq 2$ be an integer and consider $\psi(x) = \{mx\}$ on the interval $X = [0, 1)$. For example, the graph of $\psi(x)$ when $m = 5$ is illustrated in Fig. 1.1. Since $\psi^n(x) = \{m^n x\}$, we have[‡]

$$\#\operatorname{Fix}(\psi^n) = m^n - 1.$$

Thus it follows from Theorem 1.1 that

$$m^p - 1 = m - 1 + \#\operatorname{Per}_p(\psi)$$

for every prime p, and hence we have

$$m^p \equiv m \pmod{p}.$$

This derivation of Fermat's little theorem is due to Smale (1967). Later this was rediscovered by Levine (1999) and also by Frame, Johnson and Sauerberg (2000).

An expansion in base m of a real number $x \in [0, 1)$ is also obtained by iterating $\psi(x)$. Indeed x can be expanded as follows:

$$x = \sum_{n=0}^{\infty} \frac{\chi(\psi^n(x))}{m^{n+1}},$$

where $\chi(x) = [mx]$. This is derived by repeated usage of the identity

$$mx = \chi(x) + \psi(x).$$

Clearly the expansion in base m of any periodic point of ψ has recurring digits. Solving the equation $\psi^n(x) = x$, we see that the set of all fixed and periodic points

$$\operatorname{Per}(\psi) = \left\{ \frac{k}{m^n - 1} : 0 \leq k < m^n - 1, n \geq 1 \right\}$$

is dense in the unit interval $[0, 1]$.

Example 1.3. Let $\theta \in (0, 1)$ be a constant and consider $\psi(x) = \{x + \theta\}$ on the interval $X = [0, 1)$. It is easily verified that

$$\psi^n(x) = \{x + n\theta\}.$$

If we identify the interval $[0, 1)$ with the circle S^1, the corresponding mapping on S^1 becomes a homeomorphism of S^1, which is called a rigid rotation. The orbit structure of ψ depends highly on arithmetical properties of the constant θ. We distinguish two cases as follows:

(a) If θ is rational, say p/q with $\gcd(p, q) = 1$, then $\psi^n(x) = x$ occurs if and

[‡] For a finite set E we denote the number of elements in E by $\#E$.

only if $n \equiv 0 \pmod{q}$. Moreover, in such a case $\psi^n(x) = x$ holds for any x, and hence we have $\mathrm{Per}(\psi) = \mathrm{Per}_q(\psi) = [0, 1)$.

(b) If θ is irrational, then $\mathrm{Per}(\psi) = \emptyset$ because $n\theta \notin \mathbb{Z}$ for all $n \geq 1$. In particular, every orbit $\mathrm{Orb}_\psi(x)$ is infinite. More precisely, we can show that every orbit is dense in the interval $[0, 1)$. To see this, we divide $[0, 1)$ equally into m subintervals:

$$\left[0, \frac{1}{m}\right), \left[\frac{1}{m}, \frac{2}{m}\right), \ldots, \left[\frac{m-1}{m}, 1\right).$$

The $m + 1$ points $\{x, \psi(x), \ldots, \psi^m(x)\}$ are distributed on these subintervals. Thus it follows from the pigeonhole principle that at least one subinterval must contain more than one point, say $\psi^j(x)$ and $\psi^k(x)$, where $0 \leq j < k \leq m$. Putting $\alpha = \psi^k(x) - \psi^j(x)$, we have $0 < |\alpha| < 1/m$. Since

$$\alpha \equiv (k - j)\theta \pmod 1,$$

one has $\{n\alpha\} = \{n(k - j)\theta\}$ for any $n \geq 1$, and therefore

$$\begin{aligned} \psi^{n(k-j)}(x) &= \{x + n(k - j)\theta\} \\ &= \{x + n\alpha\}. \end{aligned}$$

This shows that $\mathrm{Orb}_\psi(x)$ is dense in the interval $[0, 1)$, because m and n are arbitrary.

1.4 Invariant Properties under Iterations

Invariant properties under iterations play important role in discrete dynamical systems. Clearly the piecewise linearity and piecewise monotonicity defined in the previous section are invariant under iterations. Moreover it is easily verified that surjectivity and injectivity are also invariant under iterations.

An invariant property under iterations, say $\mathfrak{P}$, is defined as follows: "If $f, g : X \to X$ satisfy the property $\mathfrak{P}$, then the composite function $f \circ g$ also satisfies $\mathfrak{P}$".

It may be worthwhile to give attention to the left limit of composite functions. Let $\mathfrak{M}$ be the set of all piecewise monotonically increasing functions which maps the interval $X = [0, 1]$ into itself. For any $f \in \mathfrak{M}$ we define

$$\tilde{f}(x) = \lim_{t \to x-} f(t)$$

for $x \in (0, 1]$, which we call the *left limit function* of f. We will use this notation throughout the book. Since $\tilde{f}(x) > 0$, we see that $\tilde{f}$ maps the interval $(0, 1]$ into itself. Note that $\tilde{f}$ differs from f at only finitely many points.

Lemma 1.2. *We have* $\widetilde{f \circ g} = \tilde{f} \circ \tilde{g}$ *for any* $f, g \in \mathfrak{M}$.

Proof. We put $\alpha = \tilde{g}(x)$ for $x \in (0,1]$. Since $g(t) \to \alpha-$ as $t \to x-$, one obtains

$$\begin{aligned} \widetilde{f \circ g}(x) &= \lim_{t \to x-} f \circ g(t) \\ &= \lim_{t \to \alpha-} f(t) \\ &= \tilde{f}(\alpha) \\ &= \tilde{f} \circ \tilde{g}(x). \end{aligned}$$ □

Note that Lemma 1.2 implies that $\widetilde{f^n} = \tilde{f}^n$ for any $n \geq 1$ and $f \in \mathfrak{M}$.

1.5 Topological Conjugacy

We say that $f : X \to X$ is *topologically conjugate* to $g : Y \to Y$ if there exists a homeomorphism $h : X \to Y$ satisfying

$$h \circ f(x) = g \circ h(x)$$

for any $x \in X$. If f is topologically conjugate to g, then it is easily seen that

$$h \circ f^n(x) = g^n \circ h(x)$$

for all $n \geq 1$, and hence

$$h(\mathrm{Orb}_f(x)) = \mathrm{Orb}_g(h(x))$$

holds. This means that topological properties of orbit structures for f and g are almost similar. For example, the orbit of x by f converges to a point $z \in X$ if and only if the orbit of $h(x)$ by g converges to the corresponding point $h(z) \in Y$.

Furthermore, if f, g and h are differentiable and $z \in \mathrm{Per}_m(f)$, then differentiating $h \circ f^m = g^m \circ h$ at z we get

$$h'(z) \cdot (f^m)'(z) = (g^m)'(h(z)) \cdot h'(z);$$

in addition, if $h'(z) \neq 0$, then

$$(f^m)'(z) = (g^m)'(h(z)).$$

This means that the local stability of f at a periodic point z conforms to that of g at $h(z)$.

1.6 Symbolic Dynamics

Let Σ be the set of all one-sided sequences $\{a_n\}_{n\geq 1}$ with $a_n \in \{0, 1\}$. The set Σ can also be identified as $\{0, 1\}^{\mathbb{N}}$, the set of all mappings from $\mathbb{N}$ to $\{0, 1\}$. The set Σ, endowed with the following metric d:

$$d(\{a_n\}, \{b_n\}) = \sum_{n=1}^{\infty} \frac{|a_n - b_n|}{2^n}, \tag{1.4}$$

becomes a compact, perfect and totally disconnected metric space. As a consequence, the set Σ is homeomorphic to *Cantor's ternary set.*

The mapping $\sigma : \Sigma \to \Sigma$ defined by

$$\sigma(a_1 a_2 a_3 \cdots) = a_2 a_3 a_4 \cdots$$

is called the *shift* operator and continuous on Σ. The expansion in base m stated in Example 1.2 gives an instructive example of the mapping Φ from the interval $[0, 1)$ into the discrete space $\Sigma' = \{0, 1, \ldots, m-1\}^{\mathbb{N}}$ with m symbols, where Φ must be discontinuous, because Σ' is not connected.

There is another way to define mappings on the set Σ compendiously. For example, consider the mapping $\tau : \Sigma \to \Sigma$ defined by

$$\tau(a_1 a_2 a_3 \cdots) = \varsigma(a_1)\varsigma(a_2)\varsigma(a_3)\cdots \tag{1.5}$$

with $\varsigma(0) = 01$ and $\varsigma(1) = 0$, where ς is known as the *Fibonacci substitution.* Then τ has a unique fixed point

$$0100101001001010010100100\cdots \in \Sigma$$

and every orbit by τ converges to this fixed point. To see this, note that the finite word $w_n = \varsigma^n(1)$ satisfies $w_{n+1} = w_n w_{n-1}$. For example,

$$\begin{aligned} w_0 &= 1, \\ w_1 &= 0, \\ w_2 &= 01, \\ w_3 &= 010, \\ w_4 &= 01001, \\ w_5 &= 01001010, \\ &\vdots \end{aligned}$$

Then it is easily seen that $\tau^n(a_1 a_2 a_3 \cdots) = w_n \cdots$ converges to the fixed point as $n \to \infty$ for any $a_1 a_2 a_3 \cdots \in \Sigma$.

Exercises in Chapter 1

1 Let I be a bounded closed interval in $\mathbb{R}$. Show that $\mathrm{Fix}(f) \neq \emptyset$ if $f : I \to I$ is continuous.

2 Let X be a complete metric space with metric d. A mapping $f : X \to X$ is said to be a *contraction* if there exists a constant $\lambda \in (0, 1)$ satisfying

$$d(f(x), f(y)) \leq \lambda\, d(x, y)$$

for any $x, y \in X$. Show that there exists a unique fixed point of f if $f : X \to X$ is a contraction. Show also that every orbit converges to this fixed point.

3 Let I be an interval in $\mathbb{R}$. For any $f \in C^3(I)$

$$S(f)(x) = \frac{f'''(x)}{f'(x)} - \frac{3}{2}\left(\frac{f''(x)}{f'(x)}\right)^2$$

is said to be the *Schwarzian derivative* of f. Show that

$$S(f \circ g) = S(f) \circ g \cdot (g')^2 + S(g)$$

and deduce from this that "$S(f)(x) < 0$ unless $f'(x) = 0$" is an invariant property under iterations for any $f, g : I \to I$ in $C^3(I)$.

4 Let $m \geq 2$ be an integer. Put $\psi(x) = \{mx\}$ and $h(x) = 1 - x$. Show then that

$$h \circ \psi(x) = \tilde{\psi} \circ h(x)$$

for $0 \leq x < 1$.

5 Let $f(x) = 4x(1 - x)$ and $\psi(x) = 1 - |1 - 2x|$ on the interval $X = [0, 1]$. Confirm that ψ is topologically conjugate to f with $h(x) = \sin^2(\pi x/2)$. We sometimes call ψ the *tent map*.

6 Establish that the metric (1.4) satisfies the axioms of metric and that Σ is compact.

7 Show that there exists a sequence $\{a_n\} \in \Sigma$ such that $\mathrm{Orb}_\sigma(\{a_n\})$ is dense in Σ.

8 Show that $\mathrm{Orb}_f(x)$ is a finite set if and only if there exist integers $m \geq 0$ and $n \geq 1$ satisfying

$$f^m(x) \in \mathrm{Per}_n(f).$$

9 Let d be the metric defined in (1.4) and $\tau : \Sigma \to \Sigma$ be the mapping defined in (1.5). Show then that

$$d(\tau(\{a_n\}), \tau(\{b_n\})) \leq d(\{a_n\}, \{b_n\})$$

for any $\{a_n\}, \{b_n\} \in \Sigma$ and that τ is not a contraction.

Chapter 2

Caianiello's Equations

We explain in brief Caianiello's neuronic equations describing neural nets by binary decision system. As a specific case of Caianiello's equations, Nagumo-Sato's equation gives an account of the output of a single neuron with a stable input pulse. Their equation can be converted to some discontinuous piecewise linear transformation, which has remarkable applications to Diophantine approximations. We also discuss a higher-dimensional generalization of Nagumo-Sato's equation.

2.1 Brief History

Eduardo R. Caianiello (1921–1993) was an Italian theoretical physicist, born in Naples. He made a contribution to quantum theory, cybernetics and the theory of neural networks. In 1957 he founded and directed the Institute of Theoretical Physics of the University of Naples. Caianiello's interest in cybernetics was indirectly vitalized by Enrico Fermi, who promoted a seminar on computers and cybernetics in the University of Rome. In 1968 Caianiello also founded and directed the Laboratory of Cybernetics at Arco Felice. He established relations with the People's Republic of China, India and Japan especially.

Cybernetics was propounded by Norbert Wiener (1894–1964), an American mathematician who contributed mainly to potential theory, generalized harmonic analysis and probability theory, with the view toward the unifying approach to physiology, mechanical and systems engineerings by blending control and telecommunications engineerings. To quote partly the introduction of Caianiello (1961), "...; herein lies indeed clearly the heart and scope of cybernetics, which aims at synthesis as well as analysis." Wiener stayed at Caianiello's institution in Naples for a year and a half.

Kurt Mahler (1903–1988) was born in Prussian Rhineland. He was a mathematician who contributed mainly to the theory of numbers, in particular transcendental numbers. His first mathematical paper entitled "On the translation properties of a simple class of arithmetical functions" was published as the Part Two of Wiener's paper in *J. Math. and Phys.* in 1927. Mahler had worked as an unpaid assistant to Wiener at Göttingen.

We will see later that some transcendental functions studied by Mahler are behind the back of Caianiello's neuronic equations.

2.2 Caianiello's Neuronic Equations

To understand thought-processes and mental phenomena quantitatively, like memories, learning, pattern recognition, self-organization and many others, Caianiello (1961) proposed two sets of equations, describing behavior of a "model of brain" or "thinking machine". The first set of equations, known as neuronic equations, determine the instantaneous behavior, or account for the behavior of networks with frozen connections, while the second, known as mnemonic equations, determine the long-term behavior, or account for the changes of coupling among neurons. He also emphasized that our machine does not necessarily purport to realize an anatomical model of the brain, but rather a physiological model. In this sense the basic component of the machine is simply called a "neuron".

For neuronic equations Caianiello imposed the following five assumptions on the machine:

(i) Signals can only travel unidirectionally with infinite speed from the output of a neuron to the inputs of the neurons connecting to it.
(ii) When a signal reaches a neuron, it is annihilated unless enough signals arrive with it to cause the neuron to fire a pulse, after a delay τ simultaneously in all its outputs.
(iii) All pulses are normalized to unit strength and are accounted for larger or smaller strengths by giving suitable coupling coefficients.
(iv) A neuron will fire only if the total sum of afferent pulses is greater than its threshold.
(v) All coupling coefficients and thresholds are constant (adiabatic learning approximation).

We furthermore assume that neurons are labeled from 1 to N and that time is discretized and normalized to $\tau = 1$; so we use a non-negative integer n as time variable. Taking into account the above five assumptions, Caianiello's neuronic equations may take the following form:

$$u_h(n+1) = \mathrm{H}\left(\sum_{k=1}^{N}\sum_{r=0}^{n} a_{hk}^{(r)} u_k(n-r) - s_h\right) \tag{2.1}$$

for $1 \le h \le N$, where the sequence $\{u_h(n)\}_{n\ge 0}$ is an element of $\Sigma = \{0,1\}^{\mathbb{N}}$ and

$$\mathrm{H}(x) = \begin{cases} 1 & (x \ge 0), \\ 0 & (x < 0), \end{cases}$$

is known as the Heaviside step function†, or as the unit step function. A neuron h is said to be excited or firing at time n if $u_h(n) = 1$.

If $h \ne k$, then $a_{hk}^{(0)}$ is the coupling coefficient transferring the pulse from a neuron k to a neuron h, which means a facilitation if $a_{hk}^{(0)} > 0$, or an inhibition if $a_{hk}^{(0)} < 0$; so, $a_{hk}^{(0)} \ne 0$ indicates that there is a unidirectional direct channel between the neurons k and h. If $h \ne k$ and $r \ge 1$, then $a_{hk}^{(r)} \ne 0$ means that the effect of the pulse from the neuron k lasts for some time on the neuron h after the neuron h has ceased firing. On the other hand, $a_{hh}^{(r)} \ne 0$ means that a neuron h retains the memory of its firing and it might be natural to assume that $a_{hh}^{(r)} \le 0$ in any case. The number s_h means the threshold of a neuron h so that it fires at time $n+1$ if the sum in (2.1), called the excitation at time n, is greater than or equal to s_h.

The equations (2.1) can be expressed vectorially as follows:

$$\boldsymbol{u}_{n+1} = \mathbf{H}\left(\sum_{r=0}^{n} A_r \boldsymbol{u}_{n-r} - \boldsymbol{s}\right), \tag{2.2}$$

where $A_r = (a_{ij}^{(r)})$ is a square matrix of size N, called the coupling coefficients matrix,

$$\boldsymbol{u}_n = \begin{pmatrix} u_1(n) \\ u_2(n) \\ \vdots \\ u_N(n) \end{pmatrix}, \quad \boldsymbol{s} = \begin{pmatrix} s_1 \\ s_2 \\ \vdots \\ s_N \end{pmatrix} \in \mathbb{R}^N \quad \text{and} \quad \mathbf{H}\begin{pmatrix} x_1 \\ x_2 \\ \vdots \\ x_N \end{pmatrix} = \begin{pmatrix} \mathrm{H}(x_1) \\ \mathrm{H}(x_2) \\ \vdots \\ \mathrm{H}(x_N) \end{pmatrix}.$$

Given an initial state $\boldsymbol{u}_0$ the equation (2.2) determines the vector $\boldsymbol{u}_n$ for all n.

The difficulty of solving (2.2) lies firstly in the fact that it contains a large amount of parameters A_r $(r \ge 0)$, which may force us to impose further assumptions for rigorous investigation. Secondly, in order to determine the value of $\boldsymbol{u}_{n+1}$ by (2.2) we need the $n+1$ values of $\boldsymbol{u}_0, \ldots, \boldsymbol{u}_n$. From the standpoint of dynamical systems it is desirable that there exists an N-dimensional dynamical system Φ on

† Although the assumption (iv) requires $\mathrm{H}(0) = 0$, we define $\mathrm{H}(0) = 1$ for convenience.

some compact subset of $\mathbb{R}^N$ and an initial point $\boldsymbol{z}_0 \in \mathbb{R}^N$ satisfying

$$\boldsymbol{u}_{n+1} = \mathbf{H}(\Phi^n(\boldsymbol{z}_0)) \tag{2.3}$$

for all $n \geq 0$. If so, the value of $\boldsymbol{u}_{n+1}$ would be determined by the orbit of Φ, a lot simpler than (2.2).

2.3 Nagumo-Sato's Equation

In 1972 Nagumo and Sato found an assumption guaranteeing the existence of a dynamical system Φ on $\mathbb{R}$ which satisfies the relation (2.3) for a single neuronic equation ($N = 1$). We simply write $\boldsymbol{u}_n, A_r, \boldsymbol{s}$ as u_n, a_r, s respectively. In fact, the neuronic equation for a neuron with a stable input pulse and an exponentially-weighted memory becomes

$$u_{n+1} = \mathrm{H}\left(\alpha - \beta \sum_{r=0}^{n} b^r u_{n-r} - s\right), \tag{2.4}$$

where $\alpha > 0$ means the amplitude of the stable input pulse and $a_{11}^{(r)} = -\beta b^r$ is a geometric progression with initial value $-\beta < 0$ and common ratio $0 < b < 1$. We call (2.4) Nagumo-Sato's equation. Note that the insertion of α in the equation (2.4) is equivalent to the replacement of s by $s - \alpha$.

Putting

$$x_n = 1 + \frac{b(\alpha - s)}{\beta} - \sum_{r=0}^{n} b^r u_{n-r}$$

for $n \geq 0$, we have

$$x_{n+1} - b x_n = 1 - bc - u_{n+1} \quad \text{with} \quad c = 1 - \frac{(1-b)(\alpha - s)}{\beta}.$$

Moreover, since

$$\beta(x_n - c) = \alpha - \beta \sum_{r=0}^{n} b^r u_{n-r} - s,$$

we obtain

$$u_{n+1} = \mathrm{H}(\beta(x_n - c)) = \mathrm{H}(x_n - c)$$

and $x_{n+1} = \phi(x_n)$, where ϕ is the discontinuous piecewise linear function defined by

$$\phi(x) = b(x-c) + 1 - \mathrm{H}(x-c)$$
$$= \begin{cases} 1 + b(x-c) & (x < c), \\ b(x-c) & (x \geq c). \end{cases}$$

If $c \leq 0$, then ϕ has a unique fixed point in the interval $[0, \infty)$, which is globally stable, or more precisely, every orbit by ϕ converges to this fixed point from above after finite iterations. Thus we get $u_n = 1$ for all $n \geq n_0$ for some n_0. In the same way, if $c > 1$, then ϕ has a globally stable fixed point in $(-\infty, c)$, and so $u_n = 0$ except for a finite number of n. The same conclusion holds for $c = 1$, but ϕ has no fixed points. Moreover, if $0 < c < 1$, then ϕ becomes a dynamical system on the closed unit interval $[0, 1]$ and x_n tumbles into this interval no matter what x_0 may be. Thus, without loss of generality, we can assume that $0 < c < 1$ and $0 \leq x_0 \leq 1$.

The behavior of the solution $\{u_n\}$ of (2.4) is reduced to that of some orbit of the dynamical system ϕ defined on $[0, 1]$. Indeed, it is easily verified that

$$u_{n+1} = \mathrm{H}(\Phi^n(z_0))$$

where

$$\Phi(x) = bx - c + 1 - \mathrm{H}(x) \quad \text{and} \quad z_0 = \frac{\alpha - s}{\beta} - u_0.$$

Here, Φ is a piecewise linear dynamical system on the closed interval $[-c, 1-c]$.

In this book we will investigate the orbit structure of piecewise linear functions

$$\psi_{a,b,c}(x) = \begin{cases} 1 + a(x-c) & (x < c), \\ b(x-c) & (x \geq c), \end{cases} \tag{2.5}$$

with three parameters a, b, c instead of $\phi(x)$. Owing to this slight generalization, we will be able to investigate the Hecke-Mahler series in Chapter 12.

2.4 Higher-Dimensional Analogy

The assumptions imposed by Nagumo and Sato in the previous section can be modified to deal with a neural net with N neurons.

We assume that $A_r = -B^r$ for a square matrix B of size N and put

$$\boldsymbol{x}_n = \boldsymbol{e} - B\boldsymbol{s} - \sum_{r=0}^{n} B^r \boldsymbol{u}_{n-r}$$

for $n \geq 0$, where $\boldsymbol{e}$ is the N-dimensional vector whose components are all 1. Then it is easily seen that

$$\boldsymbol{x}_{n+1} - B\boldsymbol{x}_n = (I - B)(\boldsymbol{e} - B\boldsymbol{s}) - \boldsymbol{u}_{n+1},$$

where I is the unit matrix of size N. On the other hand, it follows from (2.2) that

$$\begin{aligned}\boldsymbol{u}_{n+1} &= \mathbf{H}\left(-\sum_{r=0}^{n} B^r \boldsymbol{u}_{n-r} - \boldsymbol{s}\right) \\ &= \mathbf{H}(\boldsymbol{x}_n - \boldsymbol{e} - (I - B)\boldsymbol{s})\end{aligned}$$

and

$$\boldsymbol{x}_{n+1} = B(\boldsymbol{x}_n - \boldsymbol{c}) + \boldsymbol{e} - \mathbf{H}(\boldsymbol{x}_n - \boldsymbol{c}),$$

where $\boldsymbol{c} = \boldsymbol{e} + (I - B)\boldsymbol{s}$. Thus, putting $\boldsymbol{y}_n = \boldsymbol{x}_n - \boldsymbol{c}$, we obtain

$$\boldsymbol{u}_{n+1} = \mathbf{H}(\Phi^n(\boldsymbol{y}_0)),$$

where

$$\Phi(\boldsymbol{y}) = B\boldsymbol{y} - \boldsymbol{c} + \boldsymbol{e} - \mathbf{H}(\boldsymbol{y}) \quad \text{and} \quad \boldsymbol{y}_0 = -\boldsymbol{s} - \boldsymbol{u}_0. \tag{2.6}$$

Here, Φ is a piecewise linear transformation defined on $\mathbb{R}^N$.

Assume now that the matrix B is invertible and the linear transformation $\boldsymbol{y} = B\boldsymbol{x}$ is a contraction with respect to the maximum norm $\|\cdot\|_\infty$ defined by

$$\|\boldsymbol{x}\|_\infty = \max_{1 \leq i \leq N} |x_i| \quad \text{where} \quad \boldsymbol{x} = \begin{pmatrix} x_1 \\ x_2 \\ \vdots \\ x_N \end{pmatrix} \in \mathbb{R}^N;$$

that is, there exists a constant $0 < \lambda < 1$ satisfying

$$\|B\boldsymbol{x}\|_\infty \leq \lambda \|\boldsymbol{x}\|_\infty$$

for any $\boldsymbol{x}$. The mapping Φ in (2.6) may become a dynamical system on some sufficiently large closed hypercube in $\mathbb{R}^N$. Also it becomes one-to-one on any N-dimensional closed hypercube Q whose side has length less than $1/\lambda$ and are parallel to coordinate axes. To see this, suppose that $\Phi(\boldsymbol{x}) = \Phi(\boldsymbol{y})$ for $\boldsymbol{x}, \boldsymbol{y} \in Q$ with

$$\|\boldsymbol{x} - \boldsymbol{y}\|_\infty < \frac{1}{\lambda}.$$

If $\mathbf{H}(\boldsymbol{x}) \neq \mathbf{H}(\boldsymbol{y})$, then

$$\begin{aligned} 1 &\leq \|\mathbf{H}(\boldsymbol{x}) - \mathbf{H}(\boldsymbol{y})\|_\infty \\ &= \|B(\boldsymbol{x} - \boldsymbol{y})\|_\infty \\ &\leq \lambda \|\boldsymbol{x} - \boldsymbol{y}\|_\infty, \end{aligned}$$

a contradiction. Hence $\mathbf{H}(\boldsymbol{x}) = \mathbf{H}(\boldsymbol{y})$, and so $B\boldsymbol{x} = B\boldsymbol{y}$. Therefore $\boldsymbol{x} = \boldsymbol{y}$, because B is invertible by the assumption.

Moreover, if

$$\boldsymbol{c} = \begin{pmatrix} c_1 \\ c_2 \\ \vdots \\ c_N \end{pmatrix} \in \mathbb{R}^N$$

satisfies $0 < c_i < 1$ for all $1 \leq i \leq N$, then the mapping Φ has no fixed points in $\mathbb{R}^N$. To see this, suppose, on the contrary, that $\boldsymbol{x}_0 \in \mathrm{Fix}(\Phi)$. Let x_i and y_i be the i-th coordinates of the vectors $\boldsymbol{x}_0$ and $B\boldsymbol{x}_0$ respectively. Take the integer k satisfying $\|\boldsymbol{x}_0\|_\infty = |x_k|$. Since

$$\begin{aligned} \boldsymbol{x}_0 &= \Phi(\boldsymbol{x}_0) \\ &= B\boldsymbol{x}_0 - \boldsymbol{c} + \boldsymbol{e} - \mathbf{H}(\boldsymbol{x}_0), \end{aligned}$$

we obtain

$$\begin{aligned} y_k &= x_k + c_k - 1 + \mathrm{H}(x_k) \\ &= \begin{cases} x_k + c_k & (x_k \geq 0), \\ x_k + c_k - 1 & (x_k < 0). \end{cases} \end{aligned}$$

Put

$$\delta = \min_{1 \leq i \leq N} \{c_i, 1 - c_i\} > 0.$$

If $x_k \geq 0$, then $y_k = x_k + c_k > 0$ and so

$$|y_k| = |x_k| + c_k \geq |x_k| + \delta.$$

On the other hand, if $x_k < 0$, then $y_k = x_k + c_k - 1 < 0$ and so

$$|y_k| = 1 - x_k - c_k = |x_k| + 1 - c_k \geq |x_k| + \delta.$$

Therefore

$$\begin{aligned}
\|\boldsymbol{x}_0\|_\infty &< \|\boldsymbol{x}_0\|_\infty + \delta \\
&= |x_k| + \delta \\
&\leq |y_k| \\
&\leq \|B\boldsymbol{x}_0\|_\infty \\
&\leq \lambda \|\boldsymbol{x}_0\|_\infty \\
&\leq \|\boldsymbol{x}_0\|_\infty,
\end{aligned}$$

a contradiction. Hence we have $\mathrm{Fix}(\Phi) = \emptyset$, as required.

It seems that the switching element $\mathbf{H}(\boldsymbol{y})$ in (2.6) makes the investigation as dynamical systems still hard in general.

Chapter 3

Rotation Numbers

We introduce two very important mathematical notions to our discontinuous piecewise monotone dynamical systems: one is the Poincaré's rotation numbers and the other is the Milnor-Thurston's invariant coordinates. We also give a close relationship between them.

3.1 Set $\mathfrak{B}$

We deal with not only piecewise linear functions but also a bit more general functions. For an arbitrarily fixed constant $c \in (0, 1)$ let $\mathfrak{B} = \mathfrak{B}(c)$ be the set of all functions f defined on the unit interval $I = [0, 1]$ satisfying the following two conditions:

(1) $f(x)$ is continuous and strictly monotonically increasing on each interval $[0, c)$ and $[c, 1]$.

(2) $f(c-) = 1$, $f(c) = 0$ and $f(1) < f(0)$.

Note that every $f \in \mathfrak{B}$ is one-to-one but not onto, piecewise monotonically increasing and right-continuous. Moreover the right-continuity is an invariant property under iterations of f. Other case in which $f(c-) < 1$, $f(c) > 0$ can be easily reduced to our case by taking a suitable subinterval and an affine transformation.

Since $f(I) \subset [0, 1)$, f defines a discrete dynamical system on the interval $J = [0, 1)$. Every finite orbit of f must be periodic, because f is one-to-one. Of course, $\mathfrak{B}$ contains the piecewise linear function $\psi_{a,b,c}$ defined in (2.5) for suitable a, b.

For any $f \in \mathfrak{B}(c)$ we define the *reflection* of f by

$$f^*(x) = 1 - \tilde{f}(1 - x)$$

for $0 \leq x < 1$ and extend it to $x = 1$ continuously. It is easily verified that f^* belongs to $\mathfrak{B}(1-c)$. Note that $f^{**} = f$ and that

$$h \circ f^* = \tilde{f} \circ h$$

holds with $h(x) = 1 - x$; namely, f^* is topologically conjugate to $\tilde{f}$ on the unit interval I.

3.2 Discontinuity Points of f^n

Let $f \in \mathfrak{B}$. Note that f^n is continuous at $x = 0$, because f^n is right-continuous. For any discontinuity point x_0 of f^n in the interval $(0, 1]$, we have

$$\widetilde{f^n}(x_0) \neq f^n(x_0) = f^n(x_0+).$$

Such points can be characterized as follows:

Lemma 3.1. *A point $x_0 \in (0, 1]$ is a discontinuity point of f^n if and only if*

$$f^m(x_0) = c$$

for some integer $m \in [0, n)$.

Proof. If $f^m(x_0) \neq c$ for any $0 \leq m < n$, then clearly f^n is continuous at x_0. This contraposition shows the "only if" part of the lemma.

We next show that any point $x_0 \in (0, 1]$ satisfying $f^m(x_0) = c$ is a discontinuity point of f^n for all $n > m$. Without loss of generality, we can assume that $f^k(x_0) \neq c$ for $0 \leq k < m$; consequently f^m is continuous at x_0. By Lemma 1.2,

$$\begin{aligned}\widetilde{f^{m+1}}(x_0) &= \tilde{f} \circ f^m(x_0)\\ &= \tilde{f}(c) = 1\end{aligned}$$

and

$$f^{m+1}(x_0) = f(c) = 0.$$

This means that x_0 is a discontinuity point of f^{m+1}. Suppose now that

$$\widetilde{f^n}(x_0) = f^n(x_0)$$

for some $n > m + 1$. Then there exists an integer $\ell \in [m + 1, n)$ satisfying

$$\widetilde{f^\ell}(x_0) \neq f^\ell(x_0) \quad \text{and} \quad \widetilde{f^{\ell+1}}(x_0) = f^{\ell+1}(x_0).$$

Put $u = \widetilde{f^\ell}(x_0)$ and $v = f^\ell(x_0)$ for brevity. Since

$$\tilde{f}(u) = \widetilde{f^{\ell+1}}(x_0)$$
$$= f^{\ell+1}(x_0) = f(v)$$

by Lemma 1.2, it follows that $\tilde{f}(u) < 1$ and so $u \neq c$. We thus have $f(u) = f(v)$, which implies $u = v$. This contradiction completes the proof. □

A point $x \in I$ satisfying

$$f^n(x) = c, \quad n \geq 1$$

is said to be the *inverse image* of c by f^n and denoted by $f^{-n}(c)$ or simply by c_{-n}. Such a point is unique if it exists, because f is one-to-one. We put $c_0 = c$ and treat it as an inverse image by f^0 exceptionally. The above lemma states that the set of all discontinuity points of f^n consists of the set of all positive inverse images of c by $f^0, \ldots, f^{n-1}$. It can happen that the continuity point 0 becomes an inverse image of c. Indeed, this is the case in which 0 is a periodic point of f and such functions in $\mathfrak{B}$ will be classified as the set $\mathfrak{E}^0(p/q)$ in Section 4.5. Note that

$$f^n(x) = \tilde{f}^n(x)$$

holds for all $n \geq 1$ if $x \in (0, 1]$ is none of the inverse images of c. This property may be used in the proof of Exercise 1 in this chapter.

3.3 Itineraries

Our definitions of "address" and "itinerary" are slightly different from the original ones due to Milnor and Thurston (1988). Put $I_0 = [0, c)$ and $I_1 = [c, 1]$, where c is the constant in the definition of $\mathfrak{B}(c)$. For any $x \in I = [0, 1]$ we define

$$\epsilon(x) = \begin{cases} 0 & (x \in I_0), \\ 1 & (x \in I_1), \end{cases}$$

which we call the *address* of x. Note that

$$\epsilon(x) = \mathrm{H}(x - c)$$

for $x \in I$ where $\mathrm{H}(x)$ is the Heaviside step function. By the *itinerary* $\eta(x)$ is meant the sequence of addresses:

$$\eta(x) = (\epsilon_0(x), \epsilon_1(x), \epsilon_2(x), \ldots, \epsilon_n(x), \ldots) \in \{0, 1\}^{\mathbb{N}}$$

where

$$\epsilon_n(x) = \epsilon \circ f^n(x).$$

We will see later that the set of all itineraries $\eta([0, 1])$ is a fairly flimsy subset of Σ. Note that

$$\eta \circ f = \sigma \circ \eta$$

where σ is the shift.

We introduce the order "$<$" in Σ lexicographically; that is, $\zeta < \eta$ if there exists an integer $k \geq 1$ satisfying $\zeta_i = \eta_i$ for any $1 \leq i < k$, $\zeta_k = 0$ and $\eta_k = 1$, where $\zeta = (\zeta_i)$ and $\eta = (\eta_i)$.

Theorem 3.1. *If $x < y$, then $\eta(x) \preceq \eta(y)$. In other words, η is monotonically increasing.*

Proof. Suppose that $\epsilon_i(x) = \epsilon_i(y)$ for $0 \leq i \leq k$. Since $f^i(x)$ and $f^i(y)$ lie in the same subinterval I_0 or I_1 for each $0 \leq i \leq k$, we have $f^{k+1}(x) < f^{k+1}(y)$. This implies that $\epsilon_{k+1}(x) \leq \epsilon_{k+1}(y)$. Repeating this argument if $\epsilon_{k+1}(x) = \epsilon_{k+1}(y)$, we get $\eta(x) \preceq \eta(y)$. □

Since both f^n and ϵ are right-continuous and since f^n is piecewise monotonically increasing, it is clear that η is also right-continuous. Moreover x_0 is a discontinuity point of η if and only if $f^m(x_0) = c$ for some $m \geq 0$. If $\eta(x)$ is continuous on some interval $J_0 \subset [0, 1]$, then $\eta(x)$ must take a constant value on J_0. We thus conclude that $\eta(x)$ is a staircase map and it jumps only at inverse images of c.

3.4 Rotation Numbers

For homeomorphisms of the circle H. Poincaré introduced the rotation numbers using their "lifts" defined on $\mathbb{R}$. The rotation numbers, being invariant under topological conjugacy, are used for the topological classification. Roughly speaking, the rotation number means the average rotation angle of orbits. For the details of homeomorphisms of the circle, see, for example, §1.14 in Devaney's book(1989).

In this section we apply this method to our *discontinuous* functions f in $\mathfrak{B}(c)$. We first define

$$F(x) = f(\{x\}) + [x] + \epsilon_0(\{x\}) \tag{3.1}$$

for $x \in \mathbb{R}$, which we call a *lift* of f. Clearly $F(x)$ is continuous on every interval

$[k, k+1)$, $k \in \mathbb{Z}$ and satisfies

$$\{F(x)\} = f(\{x\})$$

for any $x \in \mathbb{R}$. Since

$$F(0-) = f(1) < f(0) = F(0)$$

and since

$$F(x+k) = F(x) + k$$

for any $x \in \mathbb{R}$ and $k \in \mathbb{Z}$, $F(x)$ is not continuous at every integral point. The last formula shows that

$$\phi_1(x) = F(x) - x$$

is a periodic function with period 1. Moreover $F(x)$ is a strictly monotonically increasing function on $\mathbb{R}$ so that we can treat F as a discontinuous dynamical system on $\mathbb{R}$.

The following concerns some basic properties of F^n.

Lemma 3.2. *For any $n \geq 1$ and any $x \in \mathbb{R}$, we have*

(i) $\{F^n(x)\} = f^n(\{x\})$,
(ii) $F^n(x+1) = F^n(x) + 1$,
(iii) $\phi_n(x) = F^n(x) - x$ *is right-continuous and periodic with period* 1,
(iv) $|F^n(x) - F^n(y)| \leq 1$ *if* $|x - y| \leq 1$.

Proof. (i) is straightforward. (ii) is true when $n = 1$. Suppose next that (ii) holds when $n = m$. Then

$$\begin{aligned} F^{m+1}(x+1) &= F(F^m(x) + 1) \\ &= F \circ F^m(x) + 1 \\ &= F^{m+1}(x) + 1. \end{aligned}$$

(iii) follows from (ii). In order to show (iv), suppose that $F^n(x) > F^n(y) + 1$. Then $F^n(x) > F^n(y+1)$ by (ii), and thus $x > y + 1$, because F^n is strictly monotonically increasing. This contraposition implies (iv). □

Theorem 3.2. *For any $f \in \mathfrak{B}$ the limit*

$$\rho(f) = \lim_{n \to \infty} \frac{1}{n} \sum_{k=0}^{n-1} \epsilon_k(x)$$

exists and is independent of the choice of $x \in I$. We call $\rho(f)$ the rotation number of f.

Proof. It follows from (iv) of Lemma 3.2 that

$$\begin{aligned}|\phi_n(x) - \phi_n(y)| &\leq |F^n(x) - F^n(y)| + |x - y| \\ &< 2\end{aligned}$$

for any $x, y \in [0, 1)$, and hence

$$\sup_{0 \leq x < 1} \phi_n(x) - \inf_{0 \leq x < 1} \phi_n(x) \leq 2$$

for any $n \geq 1$. Let α_n be the infimum in the left-hand side. Then

$$\alpha_n \leq \phi_n(x) \leq \alpha_n + 2$$

holds for any $x \in \mathbb{R}$ by (iii). Assigning m values $x = 0, F^n(0), \ldots, F^{(m-1)n}(0)$ to the last inequalities and adding them, we get

$$\alpha_n m \leq F^{mn}(0) \leq (\alpha_n + 2)m,$$

and so,

$$\frac{\alpha_n}{n} \leq \frac{F^{mn}(0)}{mn} \leq \frac{\alpha_n + 2}{n}.$$

Combining these with the inequalities when $m = 1$, we obtain

$$\left| \frac{F^{mn}(0)}{mn} - \frac{F^n(0)}{n} \right| \leq \frac{2}{n}.$$

Repeating the same argument by exchanging m for n, we get

$$\left| \frac{F^{mn}(0)}{mn} - \frac{F^m(0)}{m} \right| \leq \frac{2}{m}$$

so that

$$\left| \frac{F^m(0)}{m} - \frac{F^n(0)}{n} \right| \leq \frac{2}{n} + \frac{2}{m}.$$

This shows that $F^n(0)/n$ is a Cauchy sequence. Let ρ be the limit. For any $x \in \mathbb{R}$ it follows from (iv) of Lemma 3.2 that

$$\left| F^n(x) - F^n([x]) \right| \leq 1,$$

and hence

$$|F^n(x) - F^n(0)| \leq [x] + 1 \tag{3.2}$$

by (ii) of Lemma 3.2. This implies that the sequence $F^n(x)/n$ converges to the same limit ρ. In other words, the limit is independent of the choice of $x \in \mathbb{R}$.

On the other hand, one gets

$$[F(x)] = [x] + \epsilon_0(\{x\})$$

by (3.1) and generally

$$[F^n(x)] = [x] + \epsilon_0(\{x\}) + \epsilon_0(\{F(x)\}) + \cdots + \epsilon_0(\{F^{n-1}(x)\}).$$

Noticing that

$$\epsilon_0(\{F^k(x)\}) = \epsilon_0 \circ f^k(\{x\}) = \epsilon_k(\{x\})$$

by (i) of Lemma 3.2, we have

$$\left| \frac{F^n(x)}{n} - \frac{1}{n} \sum_{k=0}^{n-1} \epsilon_k(\{x\}) \right| \leq \frac{[x]+1}{n}, \tag{3.3}$$

which completes the proof. □

We remark that our rotation number $\rho(f)$ satisfies $\rho(f) \in [0,1]$, while that of a homeomorphism on the circle is defined as the fractional part of the limit.

Example 3.1. Let f be the function illustrated in Fig. 3.1. Since f has fixed points in $I_1 = [c, 1]$, we have immediately $\rho(f) = 1$. However we will see later that there are no piecewise linear functions $\psi_{a,b,c}$ in $\mathfrak{B}(c)$ satisfying $\rho(\psi_{a,b,c}) = 0$ or 1.

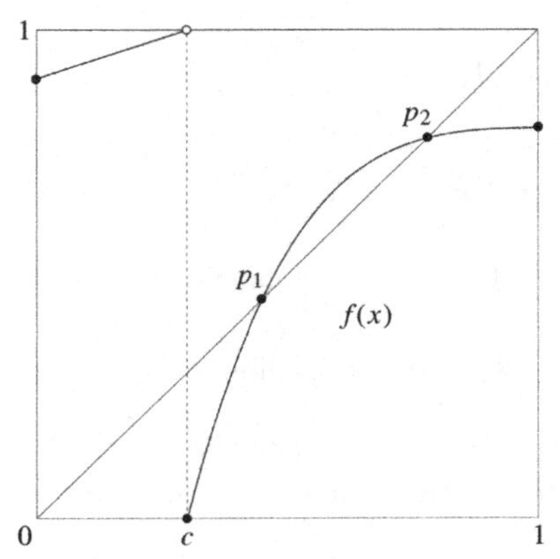

Fig. 3.1 An example of $f \in \mathfrak{B}$ with $\rho(f) = 1$. Notice that f has two fixed points p_1 and p_2, where p_1 is unstable but p_2 is stable.

The rotation number is the average of itineraries. In the point of view of neuron models, the rotation number $\rho(\psi_{a,a,c})$ when $a = b$ is known as the *average firing rate* or the *excitation number* of a neuron.

Theorem 3.3. *Let $f \in \mathfrak{B}(c)$ and $g \in \mathfrak{B}(c')$ with $c, c' \in (0, 1)$. If f is topologically conjugate to g, then $\rho(f) = \rho(g)$.*

Proof. Suppose that $h \circ f = g \circ h$ for some homeomorphism $h : [0, 1] \to [0, 1]$. Since

$$h(0) = h \circ f(c) = g \circ h(c) < 1,$$

one has $h(0) = 0$ and $h(c) = c'$. Hence $x < c$ if and only if $h(x) < c'$, which shows that $\rho(f) = \rho(g)$, as required. □

Unlike homeomorphisms of the circle, $\mathrm{Per}(f) = \emptyset$ can happen even if $\rho(f)$ is rational. With respect to this we have the following

Theorem 3.4. *Let $f \in \mathfrak{B}$. Suppose that $\rho(f) = p/q$ with $\gcd(p, q) = 1$. Then either*

$$\mathrm{Per}_q(f) \neq \emptyset \quad \text{or} \quad f^q(1) = 0$$

holds.

Proof. Put $G(x) = F^q(x) - p$ for brevity. Then $G(x)$ is a right-continuous and strictly monotonically increasing function defined on $\mathbb{R}$. It is also easily verified from Lemma 3.2 that

(i) $G^n(x + 1) = G^n(x) + 1$,

(ii) $G^n(x) = F^{nq}(x) - np$,

(iii) $\{G^n(x)\} = \{F^{nq}(x)\} = f^{nq}(\{x\})$

for any $x \in \mathbb{R}$ and all $n \geq 1$. Moreover it can be seen from (ii) that $G^n(x)/n$ tends to 0 as $n \to \infty$.

We now distinguish three cases according to the sign of $G(0)$ as follows:

(a) If $G(0) = 0$, then $f^q(0) = 0$ by (iii), and hence $0 \in \mathrm{Per}_d(f)$ for some divisor d of q. However this means that $\rho(f) = d'/d$ for some d'. We thus have $d = q$.

(b) If $G(0) < 0$, then $x_n = G^n(0)$ forms a monotonically decreasing negative sequence. We will show that $-1 < x_n < 0$ for all $n \geq 1$. Suppose, on the contrary, that $x_k \le -1$ for some $k \ge 1$. Then

$$\begin{aligned} x_{2k} &= G^k(x_k) \\ &\le G^k(-1) \\ &= G^k(0) - 1 \\ &= x_k - 1 \le -2 \end{aligned}$$

and generally we can show that $x_{nk} \le -n$ for all $n \ge 1$; and so,

$$\frac{x_{nk}}{nk} \le -\frac{1}{k}.$$

This is a contradiction, because x_m/m tends to 0 as $m \to \infty$. Hence $\{x_n\}$ is bounded and converges to some $\alpha \in [-1, 0)$. Since $G(x)$ is right-continuous, we have $\alpha \in \mathrm{Fix}(G)$ and

$$\{\alpha\} = \{G(\alpha)\} = f^q(\{\alpha\})$$

by (iii). Therefore $\{\alpha\} \in \mathrm{Per}_d(f)$ for some divisor d of q and we get $d = q$ by the same reason as (a).

(c) If $G(0) > 0$, then $y_n = G^n(0)$ forms a strictly monotonically increasing sequence and satisfies $0 < y_n < 1$ by the similar argument as (b). Let $\beta \in (0, 1]$ be the limit of y_n. We then have $G(\beta-) = \beta$ and $x < G(x) < \beta$ in a left-neighborhood of β. Letting $x \to \beta-$ in the formula $\{G(x)\} = f^q(\{x\})$ we get

$$\beta = G(\beta-) = f^q(\beta-).$$

If $f^q(x)$ is continuous at β, then $\beta \in \mathrm{Per}_d(f)$ for some divisor d of q and we have $d = q$ similarly. Otherwise, $\beta = c_{-k}$ for some $0 \le k < q$ by Lemma 3.1. Since $c_{-k} \in \mathrm{Fix}(\tilde{f}^q)$, we have $1 \in \mathrm{Fix}(\tilde{f}^q)$ and $1 \in \mathrm{Per}_d(\tilde{f})$ for some divisor d of q. Hence $0 \in \mathrm{Per}_d(f^*)$. On the other hand, since

$$\rho(f^*) = 1 - \frac{p}{q},$$

we have $d = q$ (see Exercise 4 in this chapter). As mentioned at the end of Section 3.2, if 1 were none of the inverse images of c, then $1 > f^q(1) = \tilde{f}^q(1) = 1$, a contradiction. Hence $1 = c_{-k}$ for some $k \ge 1$. Note that such an integer k is unique. Since $1 \ne c_{-j}$ for any $0 \le j < k$, it follows from Lemma 3.1 that $f^k(x)$, as well as $\tilde{f}^k(x)$, is continuous at $x = 1$. Thus $\tilde{f}^k(1) = f^k(1) = c$ and so $\tilde{f}^{k+1}(1) = \tilde{f}(c) = 1$. This implies that $k + 1 = \ell q$ for some integer $\ell \ge 1$. If $\ell \ge 2$, then $q \le \ell q - 1$ and so $1 > f^q(1) = \tilde{f}^q(1) = 1$, a contradiction.

Hence we have $k = q - 1$ and therefore $f^q(1) = f^q(c_{1-q}) = f(c) = 0$, as required. □

Corollary 3.1. *If $\rho(f)$ is either 0 or 1, we then have* $\mathrm{Fix}(f) \neq \emptyset$.

3.5 Invariant Coordinates

The kneading theory due to Milnor and Thurston (1988) is a powerful tool to analyze discrete dynamical systems with positive entropy. Since our $f \in \mathfrak{B}$ is so meek as a dynamical system, we can hardly expect many things from that theory. In this section, however, we will clarify a close relation between the rotation number and the invariant coordinate of f.

For any $x \in I$ and $f \in \mathfrak{B}$ we consider the following power series in t:

$$\Theta_f(x,t) = \sum_{n=0}^{\infty} \epsilon_n(x)t^n \in \mathbb{Z}[[t]]$$

being the generating function of the itinerary $\eta(x)$. (In the original definition this is the I_1-component of the *invariant coordinate* defined as a formal power series.)

For each fixed t it follows from the arguments on $\eta(x)$ in Section 3.3 that $\Theta_f(x,t)$ is a right-continuous staircase function in x jumping only at positive inverse images of c. In particular, we have

$$\Theta_f(1-,t) - \Theta_f(0,t) = \sum_{0<x<1} (\Theta_f(x,t) - \Theta_f(x-,t)), \tag{3.4}$$

where the summation in the right-hand side extends over all different inverse images of c by f in the interval $(0, 1)$.

Theorem 3.5. *For any $x \in I$ we have*

$$\lim_{t \to 1-} (1-t)\Theta_f(x,t) = \rho(f).$$

To show this we need the following generalization of well-known Abel's limit theorem. See, for example, Ahlfors (1979).

Lemma 3.3. *Given a sequence a_n put $s_n = a_0 + \cdots + a_n$. If*

$$b_n = \frac{s_0 + \cdots + s_n}{n}$$

converges to α, then the power series

$$\phi(z) = \sum_{n=0}^{\infty} a_n z^n$$

tends to α as z approaches 1 in such a way that $|z| < 1$ and

$$\frac{|1-z|}{1-|z|}$$

remains bounded.

Proof. Making an adjustment to the constant term a_0 we can assume that $\alpha = 0$. Since $b_n = O(1)$ and $s_n = O(n)$ as $n \to \infty$, we have

$$\phi(z) = (1-z)^2 \sum_{n=0}^{\infty} n b_n z^n$$

for $|z| < 1$. For any $\varepsilon > 0$ take n_0 so large that $|b_n| < \varepsilon$ for all $n > n_0$. Then

$$|\phi(z)| \le |1-z|^2 \sum_{0 \le n \le n_0} n|b_n| + \varepsilon |1-z|^2 \sum_{n > n_0} n|z|^n.$$

When z approaches to 1 satisfying

$$\frac{|1-z|}{1-|z|} \le K$$

for some constant K, it follows that

$$\begin{aligned}\limsup_{z\to 1} |\phi(z)| &\le \varepsilon K^2 \limsup_{z\to 1} (1-|z|)^2 \sum_{n=1}^{\infty} n|z|^n \\ &\le \varepsilon K^2.\end{aligned}$$

This completes the proof, because ε is arbitrary. □

For the proof of Theorem 3.5 we apply Lemma 3.3 with $a_0 = \epsilon_0(x)$ and

$$a_n = \epsilon_n(x) - \epsilon_{n-1}(x)$$

for $n \ge 1$. Since $s_n = \epsilon_n(x)$ and b_n converges to $\rho(f)$ by Theorem 3.2, we get

$$\lim_{t\to 1-} \phi(t) = \lim_{t\to 1-} (1-t)\Theta_f(x,t) = \rho(f),$$

as required.

In Chapters 4 and 6 we will obtain the explicit expression for $\Theta_f(0,t)$, which enables us to evaluate

$$\lim_{t\to 1-} \left(\Theta_f(0,t) - \frac{\rho(f)}{1-t} \right).$$

Exercises in Chapter 3

1 Let $f \in \mathfrak{B}$ and n be a positive integer. Show that

$$f^n(x) + f^{*n}(1-x) = 1$$

for any $x \in [0,1]$ except for $c_0, \ldots, c_{1-n}$ if exists, where f^* is the reflection of f.

2 Show that $\psi_{b,a,1-c}(x)$ is the reflection of $\psi_{a,b,c}(x)$.

3 Give an example of piecewise linear function $\psi_{a,b,c} \in \mathfrak{B}$ with $\rho(\psi) = 1/2$ and $\mathrm{Per}(f) = \emptyset$.

4 Show that

$$\rho(f) + \rho(f^*) = 1$$

for any $f \in \mathfrak{B}$, where f^* is the reflection of f.

5 Let $\{a_n\}$ be a positive sequence satisfying $a_{m+n} \le a_m + a_n$ for all $m, n \ge 1$.

(i) Show that a_n/n converges as $n \to \infty$.

(ii) Show that

$$\lim_{n\to\infty} \frac{a_n}{n} = \inf_{n\ge 1} \frac{a_n}{n}.$$

(iii) Deduce Theorem 3.2 from (i).

6 For any $f \in \mathfrak{B}$ show that

$$\Theta_f(c+,t) - \Theta_f(c-,t) = \frac{1-t}{1-t^{m+1}},$$

where m is the number of different inverse images of c by f in the interval $(0,1)$. However we understand the right-hand side as $1-t$ when $m = \infty$.

7 Give an example of the power series

$$f(x) = \sum_{n=0}^{\infty} a_n x^n$$

with radius of convergence 1 such that $f(x)$ converges as $x \to 1-$ but

$$\frac{s_0 + s_1 + \cdots + s_n}{n}$$

diverges as $n \to \infty$, where $s_n = a_0 + a_1 + \cdots + a_n$.

Chapter 4

Classification of $\mathfrak{B}$

The set $\mathfrak{B}$ defined in the previous chapter is divided into infinitely many disjoint subsets as $\mathfrak{B} = \mathfrak{B}_\infty \cup \mathfrak{B}_2 \cup \mathfrak{B}_3 \cup \cdots$. It is shown that the itinerary $\eta(x)$ is periodic with period q for any $f \in \mathfrak{B}_q$ and any x in $[0, 1)$. Each subset $\mathfrak{B}_q$ is further divided into finitely many disjoint subsets $\mathfrak{E}^\phi(p/q)$, $\mathfrak{E}^*(p/q)$ and $\mathfrak{E}^0(p/q)$ for $1 \le p < q$ with $\gcd(p, q) = 1$, according to the properties of the orbits starting from 0 and 1.

4.1 Number of Subintervals

Let $J = [0, 1)$ and $f \in \mathfrak{B}$. If J_1, ..., J_m are right-open subintervals of J satisfying $\overline{J}_i \cap \overline{J}_j = \emptyset$[†] for $i \ne j$, then the image $f(J_1 \cup \cdots \cup J_m)$ consists of m or $m + 1$ disjoint right-open subintervals of J, being not contiguous mutually. The latter occurs if and only if $c \in \operatorname{Int} J_k$[‡] for some k. Hence the closure $\overline{f^n(J)}$ consists of finitely many disjoint closed subintervals, whose number we denote by $a_n(f)$. For example, $a_1(f) = 2$ for any $f \in \mathfrak{B}$. The sequence $\{a_n(f)\}$ is monotonically increasing. We have $a_{n+1}(f) = a_n(f) + 1$ if $c \in \operatorname{Int} f^n(J)$; otherwise $a_{n+1}(f) = a_n(f)$. Moreover, since[§]

$$(0, 1) \supset \operatorname{Int} f(J) \supset \operatorname{Int} f^2(J) \supset \cdots \supset \operatorname{Int} f^n(J) \supset \cdots,$$

either of the followings must occur:

(a) $c \in \operatorname{Int} f^{q-2}(J)$ but $c \notin \operatorname{Int} f^{q-1}(J)$ for some $q \ge 2$.

(b) $c \in \operatorname{Int} f^n(J)$ for any $n \ge 1$.

---

† The *closure* of a set E is denoted by $\overline{E}$.

‡ The *interior* of a set E is denoted by $\operatorname{Int} E$.

§ If $X \supset Y$, then $\operatorname{Int} X \supset \operatorname{Int} Y$.

Let $\mathfrak{B}_q$ be the set of all $f \in \mathfrak{B}$ satisfying (a), and $\mathfrak{B}_\infty$ be the set of all $f \in \mathfrak{B}$ satisfying (b). Then

$$\mathfrak{B} = \mathfrak{B}_\infty \cup \bigcup_{q=2}^{\infty} \mathfrak{B}_q$$

is a disjoint union. For example, the function f in Example 3.1 belongs to $\mathfrak{B}_\infty$.

It is easily verified that any $f \in \mathfrak{B}_q$ satisfies

$$a_n(f) = \begin{cases} n+1 & (1 \le n < q), \\ q & (n \ge q), \end{cases}$$

while any $f \in \mathfrak{B}_\infty$ satisfies $a_n(f) = n+1$ for all $n \ge 1$. Thus we have the following

Theorem 4.1. *A function $f \in \mathfrak{B}$ belongs to $\mathfrak{B}_N$ if and only if*

$$N = \limsup_{n\to\infty} a_n(f).$$

4.2 Fundamental Properties

The next properties are valid for both $f \in \mathfrak{B}_q$ and $f \in \mathfrak{B}_\infty$.

Lemma 4.1. *Given that $c \in \operatorname{Int} f^n(J)$ for some $n \ge 0$, we have*

(i) $0 < f^k(1) < f^k(0) < 1$ *for* $1 \le k \le n+1$, *and so we put* $K_k = [f^k(1), f^k(0))$,
(ii) $\overline{K_i} \cap \overline{K_j} = \emptyset$ *for* $1 \le i \ne j \le n+1$,
(iii) $c \notin \bigcup_{k=1}^{n} \overline{K_k}$,
(iv) $f^{n+1}(J) = J \setminus \bigcup_{k=1}^{n+1} K_k$.¶

Proof. We prove by induction on n. To avoid any confusion, it is better to attach "n" to each statement (i), ..., (iv). For instance, $(\text{i})_n$, etc.

When $n = 0$ (the assumption $c \in J$ always holds), $(\text{i})_0$ and $(\text{iv})_0$ clearly hold because $f(1) < f(0)$. We next suppose that $c \in \operatorname{Int} f^{n+1}(J)$ and that each statements $(\text{i})_m, \dots, (\text{iv})_m$ hold for any $m \le n$. Since

$$c \in \operatorname{Int} f^{n+1}(J) = (0,1) \setminus \bigcup_{k=1}^{n+1} \overline{K_k}$$

from $(\text{iv})_n$, $(\text{iii})_{n+1}$ holds and $f^{n+1} : \overline{K_1} \to \overline{K_{n+2}}$ is an order-preserving homeomorphism to its image; hence $f^{n+2}(1) < f^{n+2}(0)$. We also have $0 < f^{n+2}(1)$ because

¶ The set difference $X \setminus Y$ is the set of elements in X but not in Y.

$c \notin \overline{K}_{n+1}$. Thus (i)$_{n+1}$ holds.

From (iv)$_n$ we have

$$f^{n+2}(J) \cap K_k = \emptyset$$

for $1 \le k \le n+1$, because $f^{n+2}(J) \subset f^{n+1}(J)$. Moreover $f^{n+2}(J) \cap K_{n+2} = \emptyset$ because $f^{n+1}(J) \cap K_{n+1} = \emptyset$, and so,

$$f^{n+2}(J) \subset J \setminus \bigcup_{k=1}^{n+2} K_k.$$

To show the reverse inclusion relation, let x be an arbitrary point belonging to the set in the right-hand side. Since $x \in f(J)$ from (iv)$_0$, there exists a unique inverse image $y = f^{-1}(x)$ in J satisfying

$$y \notin \bigcup_{k=1}^{n+1} K_k.$$

By (iv)$_n$ one has $y \in f^{n+1}(J)$. Hence $x \in f^{n+2}(J)$ and (iv)$_{n+1}$ holds.

Finally we show (ii)$_{n+1}$. Suppose, on the contrary, that there exists some integer $s \in [1, n+1]$ satisfying

$$\overline{K}_s \cap \overline{K}_{n+2} \neq \emptyset.$$

Since

$$\begin{aligned} \overline{K}_{n+2} &= f^{n+1}(\overline{K}_1) \\ &\subset f^{n+1}(J) \\ &= J \setminus \bigcup_{k=1}^{n+1} K_k \\ &\subset J \setminus K_s, \end{aligned}$$

the right endpoint of K_s must coincide with the left endpoint of K_{n+2}. This means that $f^s(0) = f^{n+2}(1)$, and so $f^{n+2-s}(1) = 0$, contrary to (i)$_{n+1-s}$. This completes the proof. □

As an immediate consequence of Lemma 4.1, we have the following

Corollary 4.1. *Suppose that $c \in \operatorname{Int} f^n(J)$ for some $n \ge 0$. Then each right-open subinterval constituting $f^{n+1}(J)$ takes the form of $[f^r(0), f^s(1))$ with some integers r, s satisfying $0 \le r \ne s \le n+1$.*

4.3 Periodic Behavior for $f \in \mathfrak{B}_q$

For $f \in \mathfrak{B}_q$ Lemma 4.1 holds for $n = q - 2$, and we have the following

Lemma 4.2. *For any $f \in \mathfrak{B}_q$ we have*

(i) $\overline{K_k}$, $1 \leq k \leq q - 1$ *are disjoint intervals in* $(0, 1)$,

(ii) $c \in \overline{K_{q-1}}$,

(iii) $f^{q-1}(J) = J \setminus \bigcup_{k=1}^{q-1} K_k$.

Using this lemma we can show the following

Theorem 4.2. *For any $f \in \mathfrak{B}_q$ and any $x \in J$ the itinerary $\eta(x)$ is periodic with period q. Moreover the rotation number of f is p/q where*

$$p = \sum_{n=0}^{q-1} \epsilon_n(x) \tag{4.1}$$

satisfies $\gcd(p, q) = 1$. *The above sum is independent of the choice of $x \in J$; thus, p is the number of " 1 " in the string of the first q digits in any $\eta(x)$.*

Proof. There exist at least $q - 1$ inverse images of c: let us denote them by c_0, $c_{-1}, \ldots, c_{2-q}$. These points are positive and different mutually; otherwise, 0 would be a periodic point with period less than q, contrary to (i) in Lemma 4.2 (that is, $f^k(0) > 0$ for $1 \leq k < q$). Hence these $q - 1$ points divide J into q right-open subintervals, say $J_1, \ldots, J_q$, numbered in their natural order. Since each Int J_k does not contain any inverse image of c by $f^0, \ldots, f^{q-2}$, it follows from Lemma 3.1 that each restriction $f^n | J_k$ is an order-preserving homeomorphism onto its image for any $1 \leq n < q$.

We first show that every image $f(J_k)$ is contained in some J_m. Suppose, on the contrary, that there exists $0 \leq \ell \leq q - 2$ satisfying

$$c_{-\ell} \in \operatorname{Int} f(J_k).$$

If $0 \leq \ell < q - 2$, then clearly $c_{-\ell-1} \in \operatorname{Int} J_k$, contrary to the constitution of J_i's. Otherwise (that is, if $\ell = q - 2$), we have

$$\begin{aligned} c = f^{q-2}(c_{2-q}) &\in f^{q-2}(\operatorname{Int} f(J_k)) \\ &= \operatorname{Int} f^{q-1}(J_k) \\ &\subset \operatorname{Int} f^{q-1}(J), \end{aligned}$$

contrary to $f \in \mathfrak{B}_q$.

We next show that each J_m contains at least one image of J_k by f. For any J_k whose left endpoint is c_{-r} for some $0 \le r \le q-2$, $f(J_k)$ is contained in J_m whose left endpoint is 0 or $c_{-r'}$ for some $0 \le r' \le q-3$. On the other hand, for J_m whose left endpoint is c_{2-q}, its right endpoint is 1 or c_{-s} for some $0 \le s \le q-3$. Hence J_m contains $f(J_k)$ where the right endpoint of J_k is c_0 or c_{-s-1} respectively.

The above two facts shows that each J_m contains *exactly one* image $f(J_k)$, which may conduce the permutation π of $\{1, 2, ..., q\}$ by the inclusion relation

$$f(J_k) \subset J_{\pi(k)}.$$

Let p be the number of J_k's satisfying $J_k \subset [c, 1)$. Since the right endpoint of J_{q-p} is c, the order of images $f(J_k)$ is as follows:[††]

$$f(J_{q-p+1}) \prec \cdots \prec f(J_q) \prec f(J_1) \prec \cdots \prec f(J_{q-p}).$$

Hence

$$\pi(1) = p+1, ..., \pi(q-p) = q,\ \pi(q-p+1) = 1, ..., \pi(q) = p,$$

and so $\pi = \tau^p$, where τ is the permutation

$$\tau = \begin{pmatrix} 1 & 2 & \cdots & q-1 & q \\ 2 & 3 & \cdots & q & 1 \end{pmatrix}.$$

Note that $f^q(x)$ is a strictly monotonically increasing function on J, because it maps each J_k into itself. Therefore every itinerary is periodic.

By considering the orbit starting from c_{2-q} it is easily seen that π^k is not the identity for all $1 \le k < q$. Putting $\gcd(p, q) = d$, we have $p = dr$ and $q = ds$ for some positive integers r, s. Since

$$\pi^s = \tau^{ps} = \tau^{drs} = \tau^{qr}$$

is the identity, we have $s = q$, $d = 1$ and so, $\gcd(p, q) = 1$. Hence every itinerary $\eta(x)$ is periodic with period q and the sum in (4.1) is equal to p independently of $x \in J$. □

[††] For any sets $X, Y \subset \mathbb{R}$ we write $X \prec Y$ if $X \cap Y = \emptyset$ and $\sup X \le \inf Y$.

4.4 Classification of $\mathfrak{B}_q$

For any irreducible fraction p/q in $(0, 1)$ we divide the set $\mathfrak{B}_q$ into $\varphi(q)$ subclasses exclusively as follows:

$$\mathfrak{E}(p/q) = \left\{ f \in \mathfrak{B}_q : \sum_{n=0}^{q-1} \epsilon_n(0) = p \right\},$$

where $\varphi(q)$ is called *Euler's totient function*, which is the number of positive integers less than or equal to q and relatively prime to q.

The following expresses the itinerary $\eta(0)$ explicitly for any $f \in \mathfrak{E}(p/q)$.

Theorem 4.3. *For any $f \in \mathfrak{E}(p/q)$ we have*

$$\Theta_f(0, t) = \frac{Q_{p,q}(t)}{1 - t^q}$$

where

$$Q_{p,q}(t) = \sum_{\substack{1 \le k < q \\ \{kp/q\} + p/q \ge 1}} t^k.$$

Proof. As is shown in the proof of Theorem 4.2, the permutation $\pi = \tau^p$ determines the place where each J_k is mapped by f. Consequently, the order of $\{\pi^n(1)\}$ for $0 \le n < q$ coincides with that of $\{f^n(0)\}$. Hence, if we define

$$a_n = \begin{cases} 1 & (\pi^n(1) \ge q - p + 1), \\ 0 & (\pi^n(1) \le q - p), \end{cases}$$

then the finite sequence $a_0 \cdots a_{q-1}$ gives the periodic pattern of $\eta(0)$.

For any positive integer n we put $n = sq + r$, $0 \le r < q$; that is, r is the remainder when n is divided by q. Since

$$\tau^n(1) = \tau^r(1) = r + 1,$$

we see that $\pi^n(1) = \tau^{np}(1)$ is equal to the sum of 1 and the remainder when pn is divided by q. Let $pn = s'q + r'$, $0 \le r' < q$. Then the order of $f^n(0)$, $0 \le n < q$ coincides with that of $r'/q = \{np/q\}$. This implies that $\epsilon_n(0) = 1$ if and only if

$$\frac{q - p}{q} \le \left\{ \frac{p}{q} n \right\},$$

which is equivalent to the equality

$$\left[\frac{p}{q}(n + 1) \right] - \left[\frac{p}{q} n \right] = 1.$$

Therefore we have

$$\begin{aligned}\epsilon_n(0) &= \left[\frac{p}{q}(n+1)\right] - \left[\frac{p}{q}n\right] \\ &= \left[\left\{\frac{p}{q}n\right\} + \frac{p}{q}\right]\end{aligned}$$

for $0 \leq n < q$. Since both sides of the above equalities are periodic with period q, it holds for all $n \geq 0$. This completes the proof. □

The string of the first q digits of the itinerary $\eta(0)$ is called the *basic pattern* and the corresponding polynomial

$$Q_{p,q}(t) = \sum_{k=0}^{q-1} \epsilon_k(0) t^k$$

is called the *basic pattern polynomial* of $\mathfrak{E}(p/q)$. The basic pattern always begins with "0" and ends with "1". In particular, the degree of $Q_{p,q}(t)$ is $q-1$. For example, there are four basic patterns when $q = 5$:

$$00001, \quad 00101, \quad 01011 \quad \text{and} \quad 01111;$$

therefore 00111 does not occur as a basic pattern.

By Theorem 4.3 we have the following refinement of Theorem 3.5 for $f \in \mathfrak{E}(p/q)$:

$$\Theta_f(0,t) = \frac{p}{q} \cdot \frac{1}{1-t} - \frac{1}{2} + \frac{1}{2q} + O(t-1) \tag{4.2}$$

as $t \to 1$, because $Q_{p,q}(1) = p$ and

$$\begin{aligned}Q'_{p,q}(1) &= \frac{p(q-1)}{2} + \sum_{k=1}^{q-1} \left\{\frac{p}{q}k\right\} \\ &= \frac{(p+1)(q-1)}{2}.\end{aligned}$$

4.5 Classification of $\mathfrak{E}(p/q)$

For any $f \in \mathfrak{B}_q$ we have

$$c \in \overline{K}_{q-1} = [f^{q-1}(1), f^{q-1}(0)]$$

by (ii) of Lemma 4.2. According to the position of c in $\overline{K}_{q-1}$, we further divide each $\mathfrak{E}(p/q)$ into three subclasses exclusively as follows:

$$\mathfrak{E}^{\phi}(p/q) = \left\{f \in \mathfrak{E}(p/q) : c = f^{q-1}(1)\right\},$$
$$\mathfrak{E}^{*}(p/q) = \left\{f \in \mathfrak{E}(p/q) : f^{q-1}(1) < c < f^{q-1}(0)\right\},$$
$$\mathfrak{E}^{0}(p/q) = \left\{f \in \mathfrak{E}(p/q) : c = f^{q-1}(0)\right\}.$$

We are mainly concerned with two specific orbits starting from $x = 0$ and $x = 1$.

Theorem 4.4. *For any $f \in \mathfrak{E}^{*}(p/q) \cup \mathfrak{E}^{0}(p/q)$ every orbit starting from $x \in J$ twists asymptotically around some periodic orbit with period q. More precisely,*

(i) *if $f \in \mathfrak{E}^{*}(p/q)$, then $\{f^{k+nq}(0)\}$ is strictly monotonically increasing and also $\{f^{k+nq}(1)\}$ is strictly monotonically decreasing for each $0 \le k < q$,*
(ii) *if $f \in \mathfrak{E}^{0}(p/q)$, then $0 \in \mathrm{Per}_q(f)$ and $\{f^{k+nq}(1)\}$ is strictly monotonically decreasing for each $0 \le k < q$.*

Proof. Suppose that $x = 1$ is a discontinuity point of f^q. By Lemma 3.1 we have $f^n(1) = c$ for some $n < q$, and hence $n \le q - 2$, because $c \ne f^{q-1}(1)$. However this is contrary to $c \notin \overline{K}_n$ by (i) and (ii) of Lemma 4.2. Therefore $x = 1$ is a continuity point of f^q and one has $f^q(1-) < 1$. This means that the graph of the restriction $f^q \,|\, J_q$ must cross the diagonal, as well as the other restrictions. These intersection points give a periodic orbit of f with period q. □

Differently from $\mathfrak{E}^{*}(p/q)$ and $\mathfrak{E}^{0}(p/q)$, the subclass $\mathfrak{E}^{\phi}(p/q)$ contains f satisfying $\mathrm{Per}(f) = \emptyset$. One may easily find such examples among piecewise linear functions. The following theorem and Theorem 4.8 will be used in Chapter 8.

Theorem 4.5. *For any $f \in \mathfrak{E}^{\phi}(p/q)$ we have $f^n(1) < f^n(0)$ for all $n \ge 1$. Then, putting $K_n = [f^n(1), f^n(0))$, we have $K_i \cap K_j = \emptyset$ for any $i \ne j$. Moreover, if $\mathrm{Per}_q(f) = \emptyset$, then*[♯]

$$\sum_{n=1}^{\infty} |K_n| = 1.$$

Proof. The first statement is true for $1 \le n < q$ by (i) of Lemma 4.1. Since $f^q(1) = 0$, it follows that $0 < f^q(0)$ and that

$$f^q(0) < f^{2q}(0) < f^{3q}(0) < \cdots$$

♯ The length of an interval E is denoted by $|E|$. More generally, we denote by $|X|$ the one-dimensional Lebesgue measure of $X \subset \mathbb{R}$.

from consideration of the graph of the restriction $f^q|J_1$, which is strictly monotonically increasing. For any $n \geq 1$ we have

$$f^{nq}(1) = f^{(n-1)q}(0) < f^{nq}(0),$$

and therefore we can define $K_{nq} = [f^{nq}(1), f^{nq}(0))$, where $K_q, K_{2q}, K_{3q}, \ldots$ line up adjacently in J_1. Hence

$$\bigcup_{n=1}^{\infty} K_{nq} = [0, \alpha)$$

where α is the limit of the monotonically increasing sequence $f^{nq}(0)$ as $n \to \infty$. If $\mathrm{Per}_q(f) \neq \emptyset$, then

$$\alpha = \min \mathrm{Per}_q(f);$$

otherwise, α is the right endpoint of J_1.

Since $f^q(1) = 0$, we have $f^k(1) = c_{k+1-q}$ for $1 \leq k < q$. Thus the above argument on J_1 can be applied similarly to other subintervals. In particular, if $\mathrm{Per}_q(f) = \emptyset$, then the interval J is the disjoint union of K_n's. □

Concerning the itineraries $\eta \circ f(0)$ and $\eta \circ f(1)$, we have the following

Theorem 4.6.

(a) *For any* $f \in \mathfrak{E}^{\phi}(p/q)$ *we have* $\eta \circ f(0) = \eta \circ f(1)$.

(b) *For any* $f \in \mathfrak{E}^{*}(p/q) \cup \mathfrak{E}^{0}(p/q)$ *we have* $\epsilon_n \circ f(0) = \epsilon_n \circ f(1)$ *if and only if*

$$n \not\equiv -1, -2 \pmod{q}.$$

Proof. The proof of (a) is straightforward, because $c \in K_{q-1}$. So we next show (b). It follows from (i) and (ii) of Lemma 4.2 that $c \notin \overline{K_k}$ for $1 \leq k \leq q-2$, and hence we have

$$\epsilon_n \circ f(0) = \epsilon_n \circ f(1)$$

for $0 \leq n \leq q-3$. Since $f^{q-1}(1) < c \leq f^{q-1}(0)$, we also have $\epsilon_{q-2} \circ f(0) = 1$ and $\epsilon_{q-2} \circ f(1) = 0$. Finally we get

$$\epsilon_{q-1} \circ f(0) = \epsilon_q(0) = \epsilon_0(0) = 0$$

and

$$\epsilon_{q-1} \circ f(1) = \epsilon \circ f^q(1) = \epsilon \circ f^q(1-) = 1,$$

because $f^q(x)$ is continuous at $x = 1$ by Lemma 3.1. □

4.6 Basic Properties of $f \in \mathfrak{B}_\infty$

Since Lemma 4.1 holds for any $f \in \mathfrak{B}_\infty$ and all $n \geq 1$, we have immediately the following

Lemma 4.3. *For any $f \in \mathfrak{B}_\infty$ the following properties hold:*

(i) $\overline{K_1}, \overline{K_2}, \ldots$ *are disjoint closed subintervals of* $(0, 1)$.

(ii) $c \notin \overline{K_n}$ *for all* $n \geq 1$.

(iii) $f^n(J) = J \setminus \bigcup_{k=1}^n K_k$ *for all* $n \geq 1$.

4.7 Characterization of $\mathfrak{E}^\phi$ and $\mathfrak{E}^0$

Some characterizations of the subclasses derived from Lemma 4.3, and Theorems 4.4 and 4.5 are as follows:

Theorem 4.7. *Let $q \geq 2$ be an integer and $f \in \mathfrak{B}$. Then we have*

(i) $f^q(1) = 0$ *if and only if* $f \in \mathfrak{E}^\phi(p/q)$ *for some irreducible fraction* p/q,

(ii) $0 \in \mathrm{Per}_q(f)$ *if and only if* $f \in \mathfrak{E}^0(p/q)$ *for some irreducible fraction* p/q.

Proof. Suppose that $f^q(1) = 0$. Clearly $f \notin \mathfrak{B}_\infty$ by Lemma 4.3. If $f \in \mathfrak{E}^*(r/s)$ for some irreducible fraction $r/s \in (0, 1)$, it follows from (i) of Theorem 4.4 that

$$f^s(0) = f^{q+s}(1) > f^{q+2s}(1) = f^{2s}(0),$$

contrary to $f^s(0) < f^{2s}(0)$. If $f \in \mathfrak{E}^0(r/s)$, then $0 \in \mathrm{Per}_s(f)$. However this implies that $1 = f^m(0) < 1$ for some m, a contradiction. Thus $f \in \mathfrak{E}^\phi(r/s)$, so

$$f^q(1) = f^s(1) = 0.$$

Since 1 is not a periodic point, we have $q = s$, as required.

Suppose next that $0 \in \mathrm{Per}_q(f)$. Clearly $f \notin \mathfrak{B}_\infty$ by Lemma 4.3. If $f \in \mathfrak{E}^\phi(r/s)$ for some irreducible fraction $r/s \in (0, 1)$, it follows from Theorem 4.5 that $0 = f^q(0) > 0$, a contradiction. If $f \in \mathfrak{E}^*(r/s)$, then $\mathrm{Orb}_f(0)$ is an infinite set by (i) of Theorem 4.4, contrary to $0 \in \mathrm{Per}_q(f)$. Hence we have $f \in \mathfrak{E}^0(r/s)$ and $0 \in \mathrm{Per}_s(f)$, which implies that $q = s$. This completes the proof. □

Finally we give the following. Compare with Theorem 4.5.

Theorem 4.8. *For any $f \in \mathfrak{E}^0(p/q)$ we put $K_i^* = [f^i(1), f^i(0))$ for $1 \leq i < q$ and $K_{q+k}^* = [f^{q+k}(1), f^k(1))$ for all $k \geq 0$. Then*

$$K_i^* \cap K_j^* = \emptyset \quad \textit{for any} \quad i \neq j.$$

Moreover, if $\mathrm{Per}_q(f) = \{0, f(0), f^2(0), \dots, f^{q-1}(0)\}$, *then*

$$\sum_{n=1}^{\infty} |K_n^*| = 1.$$

Proof. By (i) of Lemma 4.2 we can define $K_i^* = [f^i(1), f^i(0))$ for $1 \le i < q$. Note that each right endpoint of the interval K_i^* is the periodic point with period q, because $f^q(0) = 0$. These $q-1$ points are also the division points used in the proof of Theorem 4.2, because $c = f^{q-1}(0)$. So we define $K_q^* = [f^q(1), 1)$. In the rightmost subinterval J_q, the restriction $f^q \,|\, J_q$ is strictly monotonically increasing and has a fixed point at the left endpoint of J_q. Moreover it is continuous even at $x = 1$ by Lemma 3.1 and (i) of Theorem 4.7. Thus

$$1 > f^q(1) > f^{2q}(1) > f^{3q}(1) > \cdots$$

and we can define

$$K_{kq}^* = [f^{kq}(1), f^{(k-1)q}(1)) \quad \text{for} \quad k \ge 1,$$

standing adjacently in a queue. In addition, if

$$\mathrm{Per}_q(f) = \{0, f(0), f^2(0), \dots, f^{q-1}(0)\},$$

then the sequence $f^{kq}(1)$ must tend to the left endpoint of J_q as $k \to \infty$, and hence

$$\mathrm{Int}\, J_q = \bigcup_{k=1}^{\infty} K_{kq}^*.$$

The above argument on J_q can be applied to other subintervals so that the union of all the intervals K_n^* is equal to

$$(0, 1) \setminus \{f(0), f^2(0), \dots, f^{q-1}(0)\}.$$

This completes the proof. □

To investigate the orbit structure of $f \in \mathfrak{B}_\infty$ in details we need some basic properties of the Farey series in the next chapter.

Exercises in Chapter 4

[1] Suppose that $f \in \mathfrak{B}$ has a periodic point z with period q. Show then that $\gcd(p, q) = 1$ where

$$p = \#\left\{0 \le k < q : f^k(z) \ge c\right\}.$$

[2] Let p/q be an irreducible fraction in the interval $(0, 1)$. Show that the basic pattern polynomial can be expressed as

$$Q_{p,q}(t) = \sum_{m=1}^{p-1} t^{[mq/p]} + t^{q-1}.$$

[3] Let p/q and r/s be fractions in the interval $(0, 1)$ satisfying $qr - ps = 1$. Show that

$$Q_{p+r,q+s}(t) = Q_{p,q}(t) + t^q Q_{r,s}(t).$$

[4] Let p/q be an irreducible fraction in the interval $(0, 1)$. Show that the sum of k satisfying $1 \le k < q$ and

$$\left\{\frac{p}{q}k\right\} + \frac{p}{q} \ge 1$$

is $(p + 1)(q - 1)/2$.

[5] For any irreducible fraction p/q in the interval $(0, 1)$ show that $f \in \mathfrak{E}^0(p/q)$ if and only if $f^* \in \mathfrak{E}^\phi(1 - p/q)$, where f^* is the reflection of f.

Chapter 5

Farey Series

We recall some fundamental properties of the Farey series. Readers may find them in the books of Hardy and Wright (1979) or Niven and Zuckerman (1960). Moreover several notions as Farey intervals, predecessors, tail inversion sequences and approximation sequences are introduced and investigated for later use. In particular, our upper approximation sequences derive slightly different continued fraction expansions.

5.1 Farey Series

The *Farey series* $\mathcal{F}_n$ of order n is defined as a finite series of irreducible fractions in $[0, 1]$, whose denominators are less than or equal to n, apposed in ascending order. However we write 0 and 1 as $0/1$ and $1/1$ respectively. The first several Farey series are as follows:

$$\mathcal{F}_1 = \left\{\frac{0}{1}, \frac{1}{1}\right\},$$
$$\mathcal{F}_2 = \left\{\frac{0}{1}, \frac{1}{2}, \frac{1}{1}\right\},$$
$$\mathcal{F}_3 = \left\{\frac{0}{1}, \frac{1}{3}, \frac{1}{2}, \frac{2}{3}, \frac{1}{1}\right\},$$
$$\mathcal{F}_4 = \left\{\frac{0}{1}, \frac{1}{4}, \frac{1}{3}, \frac{1}{2}, \frac{2}{3}, \frac{3}{4}, \frac{1}{1}\right\},$$
$$\mathcal{F}_5 = \left\{\frac{0}{1}, \frac{1}{5}, \frac{1}{4}, \frac{1}{3}, \frac{2}{5}, \frac{1}{2}, \frac{3}{5}, \frac{2}{3}, \frac{3}{4}, \frac{4}{5}, \frac{1}{1}\right\},$$
$$\vdots$$

In general, $\mathcal{F}_{n+1}$ is constructed from $\mathcal{F}_n$ by inserting $(p + r)/(q + s)$ between consecutive fractions $p/q, r/s$ in $\mathcal{F}_n$ if $q + s \leq n + 1$. The fraction $(p + r)/(q + s)$ is called the *mediant* of p/q and r/s. For example, $2/5$ is the mediant of $1/3$ and $1/2$ in $\mathcal{F}_3$ but it is not contained in $\mathcal{F}_4$, because its denominator exceeds 4. The

Farey series are sometimes called the Farey sequences. The number of fractions in $\mathcal{F}_n$ is

$$1 + \varphi(1) + \varphi(2) + \cdots + \varphi(n),$$

where $\varphi(n)$ is Euler's totient function.

We first give a simple basic lemma, where the denominators of given fractions are supposed to be positive.

Lemma 5.1. *For two fractions $p/q < r/s$ satisfying $qr - ps = 1$, the mediant $(p + r)/(q + s)$ is the unique fraction lying in the interval $(p/q, r/s)$ having the smallest denominator.*

Proof. Suppose that $p/q < u/v < r/s$. Then

$$\begin{aligned}\frac{1}{qs} &= \frac{r}{s} - \frac{p}{q} \\ &= \frac{r}{s} - \frac{u}{v} + \frac{u}{v} - \frac{p}{q} \\ &\geq \frac{1}{sv} + \frac{1}{qv},\end{aligned}$$

which implies that $v \geq q + s$. However, if $v = q + s$, we then have

$$p + r - \frac{1}{q} < u < p + r + \frac{1}{s}.$$

Hence $u = p + r$ and $v = q + s$ are the unique solutions having the smallest v. □

The followings are well-known properties of the Farey series.

Lemma 5.2.

(i) *For any consecutive fractions $p/q < r/s$ in $\mathcal{F}_n$, we have $qr - ps = 1$ and $q + s \geq n + 1$.*

(ii) *For any three consecutive fractions $p/q < u/v < r/s$ in $\mathcal{F}_n$, we have*

$$\frac{u}{v} = \frac{p + r}{q + s}.$$

(iii) $\bigcup_{n=1}^{\infty} \mathcal{F}_n = [0, 1] \cap \mathbb{Q}$.

The proofs of these properties are described in almost all books on elementary number theory. For example, see Hardy and Wright (1979) or Niven and Zuckerman (1960). Lemma 5.1 will be used to show (iii) of Lemma 5.2.

Let $\mathfrak{P}(p/q)$ be a proposition concerning all rational numbers in the interval $[0, 1]$. The following lemma provides a simple way to show the correctness of $\mathfrak{P}(p/q)$ for all p/q.

Lemma 5.3. *Suppose that*

(a) $\mathfrak{P}(0)$ *and* $\mathfrak{P}(1)$ *are correct,*

(b) *for any fractions* $p/q, r/s$ *in the interval* $[0, 1]$ *with* $qr - ps = 1$,

$$\mathfrak{P}\left(\frac{p+r}{q+s}\right)$$

is correct under the assumption that both $\mathfrak{P}(p/q)$ *and* $\mathfrak{P}(r/s)$ *are correct.*

Then $\mathfrak{P}(p/q)$ *is correct for all rational numbers* p/q *in the interval* $[0, 1]$.

Proof. Suppose, on the contrary, that there exists an irreducible fraction u/v for which the proposition $\mathfrak{P}(u/v)$ does not hold. Let N be the smallest integer satisfying $u/v \in \mathcal{F}_N$. By (a) we have $N \geq 2$. Then $u/v \notin \mathcal{F}_{N-1}$ and there exist consecutive fractions $p/q, r/s$ in $\mathcal{F}_{N-1}$ such that u/v is the mediant of p/q and r/s. Since $qr - ps = 1$, $\mathfrak{P}(u/v)$ is correct by the assumption (b), a contradiction. □

Let $\mathfrak{P}(p/q, r/s)$ be a proposition concerning all fractions $p/q, r/s$ in the interval $[0, 1]$ satisfying $qr - ps = 1$. The following lemma gives a variant of Lemma 5.3, though the proof is straightforward.

Lemma 5.4. *Suppose that*

(a) $\mathfrak{P}(p_0/q_0, r_0/s_0)$ *is correct for some fractions* $p_0/q_0 < r_0/s_0$ *in the interval* $[0, 1]$ *satisfying* $q_0 r_0 - p_0 s_0 = 1$,

(b) *for any fractions* $p/q, r/s$ *in the interval* $[p_0/q_0, r_0/s_0]$ *satisfying* $qr - ps = 1$, *both the propositions*

$$\mathfrak{P}\left(\frac{p}{q}, \frac{p+r}{q+s}\right) \quad \text{and} \quad \mathfrak{P}\left(\frac{p+r}{q+s}, \frac{r}{s}\right)$$

are correct under the assumption that $\mathfrak{P}(p/q, r/s)$ *is correct.*

Then $\mathfrak{P}(p/q, r/s)$ *is correct for all fractions* $p/q, r/s$ *in the interval* $[p_0/q_0, r_0/s_0]$ *satisfying* $qr - ps = 1$.

5.2 Farey Intervals

The elements in $\mathcal{F}_n$, except for $0/1$ and $1/1$, divide $J = [0, 1)$ into finitely many right-open subintervals, which we call *Farey intervals* of order n. The set of all Farey intervals of order n is denoted by $\mathcal{I}_n$. For example, $\mathcal{I}_2$ consists of $[0, 1/2)$ and $[1/2, 1)$. Clearly the number of Farey intervals in $\mathcal{I}_n$ is

$$\varphi(1) + \varphi(2) + \cdots + \varphi(n).$$

Lemma 5.5. *Let $\alpha, \beta \in J$. Then α and β belong to the same Farey interval of order n if and only if*

$$[k\alpha] = [k\beta]$$

for $1 \le k \le n$.

Proof. Suppose that $\alpha < \beta$ and that there exists some $p/q \in \mathcal{F}_n$ satisfying $\alpha < p/q \le \beta$. Then we have $[q\alpha] \le p - 1 < p \le [q\beta]$ with $q \le n$. Conversely, if $[k\alpha] < [k\beta]$ for some $1 \le k \le n$, then

$$\alpha < \frac{[k\alpha] + 1}{k} \le \frac{[k\beta]}{k} \le \beta.$$

Therefore α and β belong to different Farey intervals of order $k \le n$. □

Given integers $0 \le i_2 \le \cdots \le i_n < n$ we define

$$\langle i_2, i_3, \ldots, i_n \rangle = \{x \in \mathbb{R} : [kx] = i_k, 2 \le k \le n\}$$
$$= \bigcap_{k=2}^{n} \left[\frac{i_k}{k}, \frac{i_k + 1}{k}\right),$$

which is clearly a right-open interval unless empty. Lemma 5.5 shows that every Farey interval in $\mathcal{I}_n$ can be expressed as $\langle i_2, i_3, \ldots, i_n \rangle$ for some $0 \le i_2 \le \cdots \le i_n < n$. For example, $\mathcal{I}_4$ consists of six subintervals:

$$\langle 0, 0, 0 \rangle,\ \langle 0, 0, 1 \rangle,\ \langle 0, 1, 1 \rangle,\ \langle 1, 1, 2 \rangle,\ \langle 1, 2, 2 \rangle,\ \langle 1, 2, 3 \rangle.$$

Note that the straightforward inequality

$$|\langle i_2, i_3, \ldots, i_n \rangle| \le \frac{1}{n}$$

implies that $qs \ge n$ for any consecutive fractions $p/q < r/s$ in $\mathcal{F}_n$. Compare with Exercise 1 in this chapter.

Therefore every $x \in J$ has a unique representation[†]

$$x = \bigcap_{n=2}^{\infty} \langle [2x], [3x], \ldots, [nx] \rangle,$$

which we may write simply as $x = \langle [2x], \ldots, [nx], \ldots \rangle$.

Lemma 5.6. *For $0 \le i_2 \le \cdots \le i_n < n$, the set $\langle i_2, i_3, \ldots, i_n \rangle$ is not empty if and only if*

$$M_n = \max_{2 \le k \le n} \frac{i_k}{k} < m_n = \min_{2 \le k \le n} \frac{i_k + 1}{k}.$$

[†] We identify the number x with the singleton set having exactly one point x.

Proof. Suppose first that $x \in \langle i_2, i_3, ..., i_n \rangle$. Since $M_n \leq x$ and $x < m_n$, we have $M_n < m_n$. Conversely, if $M_n < m_n$, then $[M_n, m_n) \subset \langle i_2, i_3, ..., i_n \rangle$. □

Let E_n be the set of lattice points (k, i_k), $2 \leq k \leq n$ in the plane of Cartesian coordinate system. Then M_n is the least number $\lambda > 0$ such that E_n is located below or on the straight line $y = \lambda x$. Similarly m_n is the largest number $\lambda > 0$ such that E_n^+ is located above or on the straight line $y = \lambda x$, where E_n^+ is the set of lattice points $(k, i_k + 1)$, $2 \leq k \leq n$, being the parallel translation of E_n just $+1$ in y-coordinate. The condition $M_n < m_n$ means that the two sets E_n and E_n^+ look completely separated, if you stand at the origin.

5.3 Predecessors

Let u/v be an arbitrary irreducible fraction in the interval $(0, 1)$. Clearly $\mathcal{F}_v$ is the first Farey series containing u/v. Then there exist two fractions $p/q < r/s$ in $\mathcal{F}_{v-1}$ whose mediant is u/v. We call p/q and r/s the left- and right-predecessors of u/v respectively. The proof of the following lemma is straightforward.

Lemma 5.7.

(i) *p/q is the left-predecessor of u/v if and only if $qu - pv = 1$ and $1 \leq q < v$.*

(ii) *r/s is the right-predecessor of u/v if and only if $rv - su = 1$ and $1 \leq s < v$.*

These notions will be used in Chapter 11, where we will see that $0/1$ behaves like the predecessor of $1/1$ and vice versa.

5.4 Floor and Ceiling Functions

Some readers might be familiar with the floor and ceiling functions, used often in computer science. They will also play a fundamental role in this book.

The *floor function* $\lfloor x \rfloor$ is the largest integer not greater than x, which is nothing but the integral part $[x]$ of x. The *ceiling function* $\lceil x \rceil$ is the smallest integer not less than x. Note that $\lfloor x \rfloor + \lceil -x \rceil = 0$, $\lceil x \rceil = \widetilde{\lfloor x \rfloor} + 1$ for any $x \in \mathbb{R}$ and $\lceil n \rceil = n$ for all $n \in \mathbb{Z}$. In particular, the ceiling function could be used to force the minus sign out of the integral part function.

In Section 5.2 we gave some basic properties concerning the integral parts of multiples of a given number. The following lemma deals with the relation between the integral parts of multiples of contiguous irreducible fractions and of its mediant, and will be used in Chapter 10.

Lemma 5.8. *Let $p/q, r/s$ be fractions in the interval $[0, \infty)$ satisfying $qr - ps = 1$. Then we have*

$$\left\lfloor \frac{p+r}{q+s} k \right\rfloor = \begin{cases} \lfloor pk/q \rfloor & (1 \le k \le q), \\ p + \lfloor r(k-q)/s \rfloor & (q < k \le q+s), \end{cases} \tag{5.1}$$

and

$$\left\lceil \frac{p+r}{q+s} k \right\rceil = \begin{cases} \lceil rk/s \rceil & (1 \le k \le s), \\ r + \lceil p(k-s)/q \rceil & (s < k \le q+s). \end{cases} \tag{5.2}$$

Proof. Put $\gamma_k = \lfloor (p+r)k/(q+s) \rfloor$ for brevity. Since

$$0 < \frac{p+r}{q+s} k - \frac{p}{q} k = \frac{k}{q(q+s)} < \frac{1}{q}$$

for $1 \le k < q+s$, we have $\gamma_k = \lfloor pk/q \rfloor$. Similarly putting $\delta_k = \lceil (p+r)k/(q+s) \rceil$, we get $\delta_k = \gamma_k + 1$ for $1 \le k < q+s$ and $\delta_{q+s} = \gamma_{q+s}$. Since

$$0 < \frac{r}{s} k - \frac{p+r}{q+s} k = \frac{k}{s(q+s)} < \frac{1}{s}$$

for $1 \le k < q+s$, we also have $\delta_k = \lceil rk/s \rceil$. Therefore, for $q < k < q+s$, we obtain

$$\begin{aligned} \gamma_k &= \left\lceil \frac{r}{s} k \right\rceil - 1 \\ &= p - 1 + \left\lceil \frac{r}{s}(k-q) + \frac{1}{s} \right\rceil. \end{aligned}$$

Since $1 \le k - q < s$, we thus have

$$\begin{aligned} \left\lceil \frac{r}{s}(k-q) + \frac{1}{s} \right\rceil &= \left\lceil \frac{r}{s}(k-q) \right\rceil \\ &= \left\lfloor \frac{r}{s}(k-q) \right\rfloor + 1, \end{aligned}$$

and hence $\gamma_k = p + \lfloor r(k-q)/s \rfloor$. This shows the second half of (5.1), because it is true even if $k = q+s$. Similarly, for $s < k < q+s$, we have

$$\begin{aligned} \delta_k &= \left\lfloor \frac{p}{q} k \right\rfloor + 1 \\ &= r + 1 + \left\lfloor \frac{p}{q}(k-s) - \frac{1}{q} \right\rfloor. \end{aligned}$$

Since $1 \le k - s < q$, we thus get

$$\left[\frac{p}{q}(k-s) - \frac{1}{q}\right] = \left[\frac{p}{q}(k-s)\right]$$
$$= \left\lceil\frac{p}{q}(k-s)\right\rceil - 1.$$

This shows the second half of (5.2), because it is true even if $k = q + s$. This completes the proof. □

5.5 Tail Inversion Sequences

For any real number λ in the interval $[0, 1]$ we define

$$\mathfrak{S}(\lambda) = (a_1, a_2, a_3, \ldots) \in \{0, 1\}^{\mathbb{N}}$$

where $a_n = [\lambda(n+1)] - [\lambda n]$[‡]. Obviously $\mathfrak{S}(p/q)$ is a periodic sequence with period q for any irreducible fraction p/q in $(0, 1)$, which is the itinerary of $f(0)$ for $f \in \mathfrak{C}(p/q)$ by Theorem 4.3.

When $q \ge 2$, the tail inversion sequence of $\mathfrak{S}(p/q)$, denoted by $\mathfrak{S}^*(p/q)$, is defined as the periodic sequence with period q whose first q digits are

$$a_1, a_2, a_3, \ldots, a_{q-2}, 1 - a_{q-1}, 1 - a_q.$$

Namely the "tail" means the "last two digits" of one period[§]. As is already shown in (ii) of Theorem 4.6, the itinerary of $f(1)$ is the tail inversion sequence of $\mathfrak{S}(p/q)$ for $f \in \mathfrak{C}^*(p/q) \cup \mathfrak{C}^0(p/q)$.

As an application of Lemma 5.5, we have

Lemma 5.9. *Let $\lambda \in [0, 1)$ be a real number. Suppose that*

$$\frac{p}{q} \le \lambda < \frac{u}{v} < \frac{r}{s}$$

where $p/q, r/s, u/v$ are fractions in the interval $[0, 1)$ satisfying $qu - pv = 1$ and $rv - su = 1$. Then the first $v - 2$ digits of the sequence $\mathfrak{S}(\lambda)$ coincide with that of $\mathfrak{S}^*(r/s)$.

---

‡ When λ is an irrational number, the sequence $\mathfrak{S}(\lambda)$ is known as a standard Sturmian word. It is known that Sturmian word has exactly $n+1$ distinct contiguous subsequences of length n for all n. For example, the sequence

$$\mathfrak{S}(2^{-1/2}) = 11011011101101\ldots$$

contains just four kinds of subsequences of length 3, which are 011, 101, 110 and 111.

§ Note that the number of "1" in the first q digits is invariant under tail inversion, because $a_{q-1} = 1$ and $a_q = 0$ if $q \ge 2$. This means that the period of $\mathfrak{S}^*(p/q)$ is exactly q, because $\gcd(p, q) = 1$.

Proof. By Lemma 5.1, $[p/q, u/v)$ is a Farey interval in $\mathcal{I}_{q+v-1}$. Since λ and p/q belong to the same Farey interval of order $q+v-1$, it follows from Lemma 5.5 that $\lfloor \lambda k \rfloor = \lfloor pk/q \rfloor$ holds for any $1 \le k \le v$. Moreover, since

$$0 < \frac{u}{v}k - \frac{p}{q}k = \frac{k}{qv} < \frac{1}{q},$$

for $1 \le k < v$, we have $\lfloor uk/v \rfloor = \lfloor pk/q \rfloor$. On the other hand, for $1 \le k < v$, we get

$$0 < \frac{r}{s}k - \frac{u}{v}k = \frac{k}{sv} < \frac{1}{s},$$

which yields that $\lceil rk/s \rceil = \lceil uk/v \rceil$. Hence

$$\lfloor \lambda k \rfloor = \left\lfloor \frac{p}{q}k \right\rfloor = \left\lfloor \frac{u}{v}k \right\rfloor = \begin{cases} \left\lfloor \frac{r}{s}k \right\rfloor & (k \not\equiv 0 \pmod{s}), \\ \left\lfloor \frac{r}{s}k \right\rfloor - 1 & (k \equiv 0 \pmod{s}), \end{cases}$$

for $1 \le k < v$. Here note that $s \ge 2$, because $r/s < 1$.

For $1 \le k \le v-2$ we distinguish three cases, as follows:

(a) $k \not\equiv 0, -1 \pmod{s}$.
We have

$$\lfloor \lambda(k+1) \rfloor - \lfloor \lambda k \rfloor = \left\lfloor \frac{r}{s}(k+1) \right\rfloor - \left\lfloor \frac{r}{s}k \right\rfloor.$$

(b) $k \equiv 0 \pmod{s}$.
Since

$$\lfloor \lambda(k+1) \rfloor - \lfloor \lambda k \rfloor = \left\lfloor \frac{r}{s}(k+1) \right\rfloor - \left\lfloor \frac{r}{s}k \right\rfloor + 1,$$

we must have $\lfloor \lambda(k+1) \rfloor - \lfloor \lambda k \rfloor = 1$ and $\lfloor r(k+1)/s \rfloor - \lfloor rk/s \rfloor = 0$.

(c) $k \equiv -1 \pmod{s}$.
Since

$$\lfloor \lambda(k+1) \rfloor - \lfloor \lambda k \rfloor = \left\lfloor \frac{r}{s}(k+1) \right\rfloor - \left\lfloor \frac{r}{s}k \right\rfloor - 1,$$

we must have $\lfloor \lambda(k+1) \rfloor - \lfloor \lambda k \rfloor = 0$ and $\lfloor r(k+1)/s \rfloor - \lfloor rk/s \rfloor = 1$.

This completes the proof. □

5.6 Approximation Sequences

For any real number κ in the interval $J = [0, 1)$, we will define two sequences of irreducible fractions $\{p_n/q_n\}$, $\{r_n/s_n\}$ associated with κ as follows.

We first put $p_1/q_1 = 0/1$ and $r_1/s_1 = 1/1$, which are the endpoints of the Farey interval of order 1. Obviously we have $p_1/q_1 \le \kappa < r_1/s_1$. Let J^* be the Farey interval of the smallest order containing κ such that the right endpoint of J^* is strictly less than r_1/s_1. We then define p_2/q_2 and r_2/s_2 as the left and right endpoints of J^* respectively. Repeating these process, we get the sequences of irreducible fractions $\{p_n/q_n\}$ and $\{r_n/s_n\}$, which we call the lower and the upper approximation sequences associated with κ respectively.

Differing from the usual simple continued fraction expansions these define rational approximations even for rational numbers. The following theorem will be used in Chapter 10, whose proof is straightforward from Lemma 5.9 because $r_2/s_2 < 1$.

Theorem 5.1. *Let $\kappa \in [0, 1)$ be any real number. Then the first $s_{n+1} - 2$ digits of the sequence $\mathfrak{S}(\kappa)$ coincide with that of $\mathfrak{S}^*(r_n/s_n)$ for all $n \ge 2$.*

The lower and upper approximation sequences have the following several properties:

(i) $\{p_n/q_n\}$ is monotonically increasing and $\{r_n/s_n\}$ is strictly monotonically decreasing. These sequences converge to κ as $n \to \infty$. If κ is rational, then $p_n/q_n = \kappa$ for all sufficiently large n.

(ii) For all $n \ge 1$ we have $q_n r_n - p_n s_n = 1$, $r_n s_{n+1} - r_{n+1} s_n = 1$ and

$$\frac{p_n}{q_n} \le \kappa < \frac{r_n}{s_n}.$$

In particular, it follows from Lemma 5.7 that p_n/q_n is the left-predecessor of r_n/s_n for $n \ge 2$ and that r_n/s_n is the right-predecessor of r_{n+1}/s_{n+1} for $n \ge 1$.

(iii) For $n \ge 1$ we have

$$\begin{pmatrix} p_{n+1} \\ r_{n+1} \end{pmatrix} = \begin{pmatrix} 1 & \ell_n - 1 \\ 1 & \ell_n \end{pmatrix} \begin{pmatrix} p_n \\ r_n \end{pmatrix} \quad \text{and} \quad \begin{pmatrix} q_{n+1} \\ s_{n+1} \end{pmatrix} = \begin{pmatrix} 1 & \ell_n - 1 \\ 1 & \ell_n \end{pmatrix} \begin{pmatrix} q_n \\ s_n \end{pmatrix}, \tag{5.3}$$

where

$$\ell_n = \ell_n(\kappa) = \left[\frac{\kappa q_n - p_n}{r_n - \kappa s_n} \right] + 1.$$

The property (iii) is easily supplied by noticing that $\ell = \ell_n(\kappa)$ is the unique positive integer satisfying

$$\frac{p_n + (\ell - 1) r_n}{q_n + (\ell - 1) s_n} \le \kappa < \frac{p_n + \ell r_n}{q_n + \ell s_n}. \tag{5.4}$$

In particular, it follows that $\{s_n\}$ is strictly monotonically increasing and $\{p_n\}$, $\{q_n\}$, $\{r_n\}$ are monotonically increasing. Note that κ is rational if and only if $\ell_n(\kappa) = 1$ for all sufficiently large n.

(iv) $r_{n+1} = p_{n+1} + r_n$ and $s_{n+1} = q_{n+1} + s_n$.

(v) $r_{n+1} = (\ell_n + 1)r_n - r_{n-1}$ and $s_{n+1} = (\ell_n + 1)s_n - s_{n-1}$ with $r_0 = 1$, $s_0 = 0$.

(vi) Each r_{n+1}/s_{n+1} is the unique fraction having the smallest denominator in the interval $(\kappa, r_n/s_n)$.

This property can be verified by applying Lemma 5.1 to the Farey interval

$$\left[\frac{p_n + (\ell_n - 1)r_n}{q_n + (\ell_n - 1)s_n}, \frac{r_n}{s_n}\right).$$

We now consider the mapping $\Upsilon : [0, 1) \to \mathbb{N}^{\mathbb{N}}$ defined by

$$\Upsilon(\kappa) = (\ell_1(\kappa), \ell_2(\kappa), \ell_3(\kappa), \ldots),$$

which is obviously one-to-one. Conversely, for any sequence $\{\ell'_n\}$ in $\mathbb{N}^{\mathbb{N}}$, we can define four sequences (p'_n, q'_n, r'_n, s'_n) by (5.3) recursively with $(p'_1, q'_1, r'_1, s'_1) = (0, 1, 1, 1)$ and $\ell_n = \ell'_n$. Then it can be easily seen that $q'_n r'_n - p'_n s'_n = 1$ and $s'_n \to \infty$ as $n \to \infty$, and hence $\{p'_n/q'_n\}$ is monotonically increasing and $\{r'_n/s'_n\}$ is strictly monotonically decreasing. Therefore $[p'_n/q'_n, r'_n/s'_n]$ makes a monotonically descending sequence of closed intervals converging to some point, say κ. Thus we have

$$\kappa = \bigcap_{n=1}^{\infty} \left[\frac{p'_n}{q'_n}, \frac{r'_n}{s'_n}\right].$$

Suppose that $(p'_i, q'_i, r'_i, s'_i) = (p_i, q_i, r_i, s_i)$, $1 \le i \le m$ for some integer $m \ge 1$. Since the inequalities

$$\frac{p'_n}{q'_n} \le \kappa < \frac{r'_n}{s'_n}$$

hold for all $n \ge 1$, it follows that

$$\frac{p_i + (\ell'_i - 1)r_i}{q_i + (\ell'_i - 1)s_i} \le \kappa < \frac{p_i + \ell'_i r_i}{q_i + \ell'_i s_i}$$

and hence $\ell_i(\kappa) = \ell'_i$ by (5.4) for $1 \le i \le m$. It also follows that

$$\begin{aligned} p'_{i+1} &= p'_i + (\ell'_i - 1)r'_i \\ &= p_i + (\ell_i(\kappa) - 1)r_i = p_{i+1}, \end{aligned}$$

and similarly $q'_{i+1} = q_{i+1}$, $r'_{i+1} = r_{i+1}$ and $s'_{i+1} = s_{i+1}$. Since $(p'_1, q'_1, r'_1, s'_1) = (p_1, q_1, r_1, s_1)$, we can apply the above argument repeatedly so that $\ell_n(\kappa) = \ell'_n$ for every $n \geq 1$. This means that $\Upsilon : [0, 1) \to \mathbb{N}^{\mathbb{N}}$ is surjective. Moreover Υ becomes strictly monotonically increasing if $\mathbb{N}^{\mathbb{N}}$ is endowed with lexicographic order. For examples, $\ell_n(0) = 1$ for all $n \geq 1$ and

$$\ell_n\left(\frac{\sqrt{5}-1}{2}\right) = 2$$

for all $n \geq 1$. Note that

$$\kappa = 1 - \sum_{n=1}^{\infty} \frac{1}{s_n s_{n+1}}$$

holds for any $\kappa \in [0, 1)$.

Finally we investigate the ratio of the denominator of upper approximation sequence to that of lower approximation sequence. For any κ we put

$$\vartheta(\kappa) = \limsup_{n\to\infty} \frac{s_n}{q_n} \geq 1.$$

It is easily seen that $\vartheta(\kappa) = \infty$ if κ is rational. When κ is irrational, we have the following rough upper and lower bounds of $\vartheta(\kappa)$.

Lemma 5.10. *Let κ be irrational. Suppose that the number of digit "1" in the sequence $\Upsilon(\kappa)$ is finite. Then*

$$1 + \frac{1}{\ell^*} \leq \vartheta(\kappa) \leq 1 + \frac{1}{\ell^* - \varpi},$$

where $\ell^* = \liminf_{n\to\infty} \ell_n(\kappa) \geq 2$ and $\varpi = (3 - \sqrt{5})/2$.

Proof. It follows from the properties (iv) and (v) that

$$\begin{aligned} \frac{s_{n+1}}{q_{n+1}} &= 1 + \frac{s_n}{s_{n+1} - s_n} \\ &= 1 + \frac{1}{\ell_n - s_{n-1}/s_n} \end{aligned} \tag{5.5}$$

for all $n \geq 1$, where $\ell_n = \ell_n(\kappa)$. The lower bound follows easily from (5.5). To show the upper bound we consider the function of n variables $x_1, x_2, \ldots, x_n \geq 2$ defined by

$$[x_1, x_2, \ldots, x_n] = x_1 - \cfrac{1}{x_2 - \cfrac{1}{x_3 - \cfrac{1}{\ddots - \cfrac{1}{x_n}}}}.$$

Since it is easily seen that $[x_1, ..., x_n] > 1$ by induction on n, the above function is well-defined for any $x_i \geq 2$. Moreover $[x_1, ..., x_n]$ is strictly monotonically increasing with respect to each x_i and satisfies $[x_1, ..., x_{n-1}, \infty] = [x_1, ..., x_{n-1}]$.

Now take a sufficiently large n_0 satisfying $\ell_n \geq 2$ for all $n > n_0$. Since

$$\frac{s_{n+1}}{s_n} = \ell_n + 1 - \frac{s_{n-1}}{s_n} = [\ell_n + 1, \ell_{n-1} + 1, ..., \ell_1 + 1],$$

we have

$$\frac{s_{n+1}}{s_n} \geq [\underbrace{3, ..., 3}_{n-n_0 \text{ times}}, [\underbrace{2, ..., 2}_{n_0 \text{ times}}]] = \left[\underbrace{3, ..., 3}_{n-n_0-1 \text{ times}}, 2 + \frac{1}{n_0 + 1}\right]$$

for all $n > n_0$. Obviously the last expression converges monotonically to $1/\varpi$ as $n \to \infty$, which implies the desired upper bound of $\vartheta(\kappa)$ from (5.5). □

Lemma 5.11. *Let κ be irrational. Suppose that the number of digit "1" in the sequence $\Upsilon(\kappa)$ is infinite. If $\Upsilon(\kappa)$ contains infinitely many strings consisting of consecutive 1's of length d, while it contains only a finite number of such strings of length $d + 1$, then*

$$d + 1 \leq \vartheta(\kappa) \leq d + 2 - \frac{1}{d + 3}.$$

On the other hand, if $\Upsilon(\kappa)$ contains a string of consecutive 1's of arbitrary length, then we have $\vartheta(\kappa) = \infty$.

Proof. We write $\ell_n(\kappa)$ as ℓ_n for brevity. If $\ell_{n-1} \geq 2$ and $\ell_n = \cdots = \ell_{n+d-1} = 1$, then by definition,

$$1 \leq \frac{s_n}{q_n} < 2, \quad \frac{s_{n+1}}{q_{n+1}} = 1 + \frac{s_n}{q_n}, \quad ..., \quad \frac{s_{n+d}}{q_{n+d}} = 1 + \frac{s_{n+d-1}}{q_{n+d-1}}.$$

Hence $s_{n+d}/q_{n+d} = d + s_n/q_n$ belongs to the interval $[d + 1, d + 2)$. Since

$$\frac{s_{n-1}}{s_{n-2}} = [\ell_{n-2} + 1, ..., \ell_1 + 1] \geq [\underbrace{2, ..., 2}_{d \text{ times}}, 3, [\underbrace{2, ..., 2}_{n-d-3 \text{ times}}]] \geq [\underbrace{2, ..., 2}_{d+1 \text{ times}}] = \frac{d + 2}{d + 1}$$

for all sufficiently large n, we have

$$\begin{aligned}\frac{s_n}{q_n} &= 1 + \frac{1}{\ell_{n-1} - s_{n-2}/s_{n-1}} \\ &\leq 1 + \frac{1}{2 - (d+1)/(d+2)} \\ &= 2 - \frac{1}{d+3}.\end{aligned}$$

This completes the proof. □

The above lemmas imply that, if κ is irrational, then $\vartheta(\kappa) > 1$ if and only if the sequence $\ell_n(\kappa)$ diverges to ∞ as $n \to \infty$. Moreover $[\vartheta(\kappa)] - 1$ characterizes the integer "d" stated in Lemma 5.11.

5.7 Continued Fraction Expansions

Our approximation sequences are not completely unrelated to the usual continued fraction expansions. However ours are slightly different from the typical expansions.

The transformation

$$\phi(x) = \left\{\frac{1}{1-x}\right\}$$

maps the interval $J = [0, 1)$ onto itself, defining a discontinuous dynamical system on J, whose graph is illustrated in Fig. 5.1.

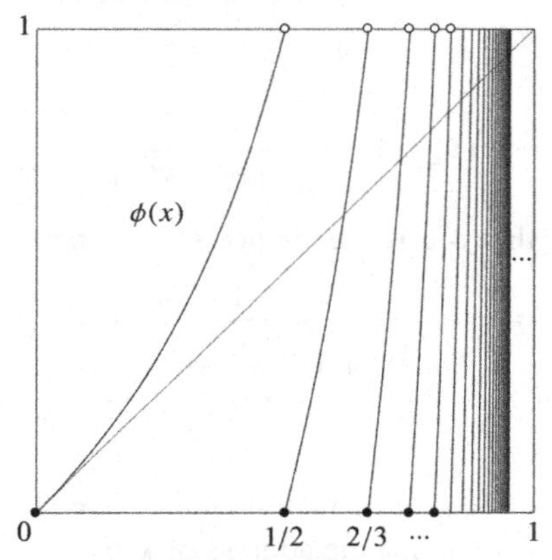

Fig. 5.1 The function $\phi(x)$ has infinitely many fixed points in $[0, 1)$. For example, the second smallest fixed point is $(\sqrt{5} - 1)/2$, the reciprocal of the golden ratio.

We then have

Lemma 5.12. *For any $\kappa \in J$ and $n \geq 0$, we have*

$$\phi^n(\kappa) = \frac{\kappa q_{n+1} - p_{n+1}}{r_n - \kappa s_n}.$$

Proof. We prove by induction on n. It is obvious when $n = 0$, because of $r_0 = 1$ and $s_0 = 0$. Suppose next that the lemma is true for some $n \geq 0$. From the property (v) in the previous section we have

$$\begin{aligned}\frac{1}{1 - \phi^n(\kappa)} &= \frac{r_n - \kappa s_n}{p_{n+1} + r_n - \kappa(q_{n+1} + s_n)} \\ &= \frac{\kappa q_{n+1} - p_{n+1}}{r_{n+1} - \kappa s_{n+1}} + 1.\end{aligned}$$

Hence the integral part of the left-hand side is $\ell_{n+1}(\kappa)$ and therefore

$$\begin{aligned}\phi^{n+1}(\kappa) &= \left\{\frac{1}{1 - \phi^n(\kappa)}\right\} \\ &= \frac{\kappa q_{n+1} - p_{n+1}}{r_{n+1} - \kappa s_{n+1}} + 1 - \ell_{n+1}(\kappa) \\ &= \frac{\kappa q_{n+2} - p_{n+2}}{r_{n+1} - \kappa s_{n+1}},\end{aligned}$$

as required. □

In the above proof we showed that

$$\frac{1}{1 - \phi^n(\kappa)} = \ell_{n+1}(\kappa) + \phi^{n+1}(\kappa),$$

and hence

$$\phi^n(\kappa) = 1 - \frac{1}{\ell_{n+1}(\kappa) + \phi^{n+1}(\kappa)} \tag{5.6}$$

for all $n \geq 0$. Applying the recurrence relation (5.6) repeatedly we can write

$$\kappa = 1 - \cfrac{1}{\ell_1 + 1 - \cfrac{1}{\ell_2 + 1 - \cfrac{1}{\ddots}}}.$$

So the mapping $\Upsilon(\kappa) = (\ell_1(\kappa), \ell_2(\kappa), \ell_3(\kappa), \ldots)$ can be regarded as the itinerary of κ with respect to ϕ if we define the address of κ by

$$\left[\frac{1}{1 - \kappa}\right].$$

It follows from the property (i) and Lemma 5.12 that κ is rational if and only if the orbit of κ by ϕ falls into 0, the smallest fixed point of ϕ.

Exercises in Chapter 5

1 For any consecutive fractions $p/q, r/s$ in $\mathcal{F}_n$ show that $q + s \geq n + 1$.

2 Show that

$$\sum_{p/q \in \mathcal{F}_n} \frac{p}{q} = \frac{1}{2} + \frac{1}{2} \sum_{k=1}^{n} \varphi(k).$$

3 Given irreducible fractions $p/q < r/s$ in the interval $[0, 1]$, we define inductively a new series $\mathcal{G}_n$ as follows:

$$\mathcal{G}_1 = \left\{ \frac{p}{q}, \frac{r}{s} \right\},$$

$$\mathcal{G}_2 = \left\{ \frac{p}{q}, \frac{p+r}{q+s}, \frac{r}{s} \right\},$$

$$\mathcal{G}_3 = \left\{ \frac{p}{q}, \frac{2p+r}{2q+s}, \frac{p+r}{q+s}, \frac{p+2r}{q+2s}, \frac{r}{s} \right\},$$

$$\vdots$$

Here $\mathcal{G}_{n+1}$ is constructed by inserting mediants between consecutive fractions in $\mathcal{G}_n$ without any condition on its denominator; however we do not reduce inserted mediants. Show then that

$$\bigcup_{n=1}^{\infty} \mathcal{G}_n = \left[\frac{p}{q}, \frac{r}{s} \right] \cap \mathbb{Q}.$$

4 Observe that $\langle 1, 2, 3, \ldots, n, \ldots \rangle = \emptyset$. So, what is

$$\bigcap_{n=2}^{\infty} \overline{\langle 1, 2, 3, \ldots, n \rangle} \ ?$$

5 Show that κ is irrational if and only if

$$\limsup_{n \to \infty} \frac{s_{n+1}}{s_n} > 1,$$

where s_n is the denominator of the upper approximation sequence associated with κ.

6 Show that

$$\frac{r_{n+1}}{s_{n+1}} = 1 - \frac{1}{[\ell_1 + 1, \ell_2 + 1, \ldots, \ell_n + 1]}$$

for all $n \geq 1$.

7 For $\kappa = \pi - 3$ check that $\ell_i(\kappa) = 1$ for $1 \leq i \leq 6$, $\ell_7(\kappa) = 16$ and $\ell_8(\kappa) = 293$. Find out the values of $3 + r_n/s_n$ for $n = 7, 8$ and 9, which are upper rational approximations to π.

8 Under the same assumptions as in Lemma 5.11 sharpen the upper bound for $\vartheta(\kappa)$ slightly by showing that

$$\vartheta(\kappa) \leq d + 1 + \frac{2\sqrt{d+1}}{\sqrt{d+1} + \sqrt{d+5}}.$$

9 Show that κ is a quadratic irrational if and only if the orbit of κ by ϕ is eventually periodic but does not fall into 0.

10 Show that $\vartheta(\kappa) = \infty$ if and only if

$$\limsup_{n \to \infty} s_n^2 \left(\frac{r_n}{s_n} - \kappa \right) = \infty.$$

Chapter 6

Further Investigation of $f \in \mathfrak{B}_\infty$

In order to determine the invariant coordinate $\Theta_f(0,t)$ for f in $\mathfrak{B}_\infty$, we continue studying the orbit structure of f. As a consequence, we show that the natural boundary of $\Theta_f(0,t)$ is the unit circle $|t| = 1$ when the rotation number of f is irrational. We also discuss an asymptotic development of $\Theta_f(0,t)$ as t tends to 1 from the left for every f in $\mathfrak{B}$.

6.1 Cases (a) and (b)

Recall that $f \in \mathfrak{B}_\infty$ has the inverse image c_{-k} of c by f^k for any $k \geq 0$. These points are all positive and different mutually; otherwise, 0 would be a periodic point, contrary to (i) of Lemma 4.3. For an arbitrarily fixed $n \geq 2$, $n-1$ points $c = c_0, c_{-1}, \ldots, c_{2-n}$ divide $J = [0,1)$ into n right-open subintervals, say $J_1, \ldots, J_n$, numbered in their natural order. Since each $\operatorname{Int} J_k$ does not contain $c_0, \ldots, c_{2-n}$, it follows from Lemma 3.1 that each restriction $f^\ell | J_k$ is an order-preserving homeomorphism to its image for $1 \leq \ell \leq n-1$.

Each closed interval $\overline{K_k}$ does not contain any inverse image of c. To see this, suppose, on the contrary, that $c_{-\ell} \in \overline{K_k}$ for some ℓ and k. Since $c_{-\ell-k} \in J$, it follows from (iii) of Lemma 4.3 that

$$c_{-\ell} \in f^k(J) = J \setminus \bigcup_{s=1}^{k} K_s.$$

Therefore $c_{-\ell} \notin K_k$, and hence $c_{-\ell} = f^k(0)$, a contradiction. Each $\overline{K_k}$ is thus contained in some subinterval J_ℓ in such a way that $J_\ell \setminus \overline{K_k}$ consists of two connected components. It is clear that $f(J_n)$ and K_1 are contiguous, as well as K_1 and $f(J_1)$. Hence

$$L = f(J_n) \cup K_1 \cup f(J_1)$$

is a right-open interval. Let m be the number of c_{-k}, $0 \leq k \leq n-2$ satisfying $c_{-k} \geq c = c_0$. We then have $J_1, \ldots, J_{n-m} \subset [0,c)$ and $J_{n-m+1}, \ldots, J_n \subset [c,1)$.

Suppose that $\operatorname{Int} f(J_k)$ contains a division point c_{-i} for some $0 \le i \le n-2$. Then $c_{-i-1} \in \operatorname{Int} J_k$, and therefore $i = n-2$. This implies that $\operatorname{Int} L$ contains at most one division point and if so, it should be c_{2-n}. We then distinguish two cases as follows:

(a) $c_{2-n} \in \operatorname{Int} L$.

Either $c_{2-n} \in \operatorname{Int} f(J_1)$ or $c_{2-n} \in \operatorname{Int} f(J_n)$ holds, because $c_{2-n} \notin \overline{K_1}$. Since the endpoints of L are division points, we have $L = J_m \cup J_{m+1}$ and each restriction $f : J_k \to J_s$ is an order-preserving homeomorphism to its image for $1 < k < n$, where $s-1$ is the remainder of $k+m-1$ divided by n. In this case the inverse image c_{2-n} is close to either $f(0)$ or $f(1)$.

(b) $c_{2-n} \notin \operatorname{Int} L$.

Differently from (a) L is some subinterval, say $J_{m'}$. If $c_{1-n} \in \operatorname{Int} J_t$ for some t, then $c_{2-n} \in \operatorname{Int} f(J_t)$. Thus J_t is mapped onto the union of two consecutive subintervals homeomorphically. Other subinterval J_k $(k \ne 1, n, t)$ is mapped onto some subinterval homeomorphically, and hence m' is m or $m+1$ according to $t \le n-m$ or not. Note that this case does not occur when $n = 2$ and $n = 3$.

In the next section we will study Case (a) in detail.

6.2 Behavior in Case (a)

Suppose first that $c_{2-n} \in \operatorname{Int} f(J_1)$. Let c^* be the smallest division point among $c_{-k}, 0 \le k \le n-2$. Then we have $J_1 = [0, c^*)$ and

$$f(0) < c_{2-n} < f(c^*-).$$

We now define

$$g(x) = \begin{cases} \dfrac{f(c^*-) - c_{2-n}}{c^*} x + c_{2-n} & (x \in J_1), \\ f(x) & (\text{otherwise}). \end{cases}$$

Obviously $g \in \mathfrak{B}(c)$ and moreover $f^{-k}(c) = g^{-k}(c)$ for $1 \le k \le n-2$, because J_1 does not contain the inverse images $f^{-k}(c)$. So we can use the notations of addresses $\epsilon_k(x)$ and inverse images c_{-k} commonly for f and g. Since $g(0) = c_{2-n}$ and $c_0, \ldots, c_{2-n}$ are mutually different, we have $0 \in \operatorname{Per}_n(g)$. Therefore it follows from (ii) of Theorem 4.7 that $g \in \mathfrak{E}^0(p/n)$, where

$$p = \sum_{k=0}^{n-1} \epsilon \circ g^k(0)$$

satisfies $\gcd(p, n) = 1$. Moreover, since

$$c_{1+k-n} = g^n(c_{1+k-n}) = g^k(0)$$

for $1 \le k \le n-1$, we have

$$\begin{aligned} p &= \sum_{k=1}^{n-1} \epsilon \circ g^k(0) \\ &= \sum_{k=1}^{n-1} \epsilon(c_{1+k-n}) = m. \end{aligned}$$

On the other hand, since $f(J_1) = [f(0), c^*-)$ does not contain $c_0, \ldots, c_{3-n}$ and since $c_{2-n} \in f(J_1)$, we have

$$\begin{aligned} \epsilon \circ f^k(0) &= \epsilon \circ f^{k-1}(c_{2-n}) \\ &= \epsilon(c_{1+k-n}) \\ &= \epsilon \circ g^k(0) \end{aligned}$$

for $1 \le k \le n-2$. Therefore it follows from Theorem 4.3 that

$$\epsilon \circ f^k(0) = \left[\frac{m}{n}(k+1)\right] - \left[\frac{m}{n}k\right]$$

for $1 \le k \le n-2$. This holds also for $k = 0$ clearly, but not for $k = n-1$ because of $f^{n-1}(0) < c$.

We next suppose that $c_{2-n} \in \operatorname{Int} f(J_n)$. Although our argument will go along in the same way as the first case, we give the details for the sake of completeness. Let c^{**} be the largest one in the division points so that $J_n = [c^{**}, 1)$ and

$$f(c^{**}) < c_{2-n} < f(1).$$

We then define

$$h(x) = \begin{cases} \dfrac{f(c^{**}) - c_{2-n}}{1 - c^{**}}(1-x) + c_{2-n} & \left(x \in \overline{J}_n\right), \\ f(x) & \text{(otherwise)}. \end{cases}$$

Obviously $h \in \mathfrak{B}(c)$ and $f^{-k}(c) = h^{-k}(c)$ for $1 \le k \le n-2$, because J_n does not contain the inverse images $f^{-k}(c)$. Since $h(1) = c_{2-n}$, we have $h^n(1) = 0$. Hence it follows from (i) of Theorem 4.7 that $h \in \mathfrak{E}^{\phi}(p/n)$, where

$$p = \sum_{k=0}^{n-1} \epsilon \circ h^k(0)$$

satisfies $\gcd(p, n) = 1$. Furthermore, since

$$c_{1+k-n} = h^{k-1}(c_{2-n}) = h^k(1)$$

for $1 \leq k \leq n-1$, we have from (i) of Theorem 4.6

$$\begin{aligned} p &= \sum_{k=1}^{n-1} \epsilon \circ h^k(0) \\ &= \sum_{k=1}^{n-1} \epsilon \circ h^k(1) \\ &= \sum_{k=1}^{n-1} \epsilon(c_{1+k-n}) = m. \end{aligned}$$

Since $f(J_n) = [f(c^{**}), f(1))$ does not contain $c_0, \ldots, c_{3-n}$ and $c_{2-n} \in f(J_n)$, we also have

$$\begin{aligned} \epsilon \circ f^k(0) &= \epsilon \circ f^k(1) \\ &= \epsilon \circ f^{k-1}(c_{2-n}) \\ &= \epsilon(c_{1+k-n}) \\ &= \epsilon \circ h^k(1) = \epsilon \circ h^k(0) \end{aligned}$$

for $1 \leq k \leq n-2$. Therefore it follows from Theorem 4.3 that

$$\epsilon \circ f^k(0) = \left[\frac{m}{n}(k+1)\right] - \left[\frac{m}{n}k\right]$$

for $1 \leq k \leq n-2$. This holds also for $k = 0$ and $k = n-1$, because $c < f^{n-1}(0)$.

So rounding up the above arguments, we have

Lemma 6.1. *Suppose that $f \in \mathfrak{B}_\infty$ and*

$$c_{2-n} \in \operatorname{Int} f(J_1) \cup \operatorname{Int} f(J_n)$$

for some $n \geq 2$. Then there exists a positive integer $m = m(n)$ with $\gcd(m, n) = 1$ such that

$$\sum_{k=0}^{s-1} \epsilon_k(0) = \left[\frac{m}{n}s\right]$$

for any $1 \leq s < n$.

6.3 Discontinuation of Case (b)

In this section we will show that Case (a) occurs for infinitely many n's.

Suppose, on the contrary, that Case (b) occurs for all $n \geq N$ for some $N \geq 4$. We consider subintervals $J_1, ..., J_N$ obtained by the division points $c_0, ..., c_{2-N}$. Obviously the leftmost subinterval $J_1 = [0, c^*)$, is invariant† for any level $n \geq N$, as well as the rightmost subinterval $J_N = [c^{**}, 1)$, where

$$0 < c^* = \min_{k \geq 0} c_{-k} < \max_{k \geq 0} c_{-k} = c^{**} < 1.$$

In other words, $J_1 \cup J_N$ does not contain any inverse images of c by f^k, $k \geq N-1$. Moreover

$$f(J_N) \cup K_1 \cup f(J_1)$$

is some subinterval, say J_M. If $J_M = J_1$, then $c^{**} = c$ and c^* is a continuity point of f. However this would imply that $c^* \in \mathrm{Per}(f)$, a contradiction. Hence $J_M \neq J_1$. By the similar reason it can be seen that $J_M \neq J_N$. Therefore we have $c^* < c_0 < c^{**}$.

We next show the following statements by induction on k:

(1) $f^k(J_M)$ is one of the subintervals $J_2, ..., J_{N-1}$.

(2) $f^k | J_M$ is an order-preserving homeomorphism onto its image.

Clearly these propositions hold for $k = 0$. Suppose next that these hold for $0 \leq k \leq s$. If f maps the subinterval $f^s(J_M)$ onto some consecutive subintervals $J_t \cup J_{t+1}$, then $c_{1-n} \in \mathrm{Int}\, f^s(J_M)$ and hence $c_{1-n-s} \in \mathrm{Int}\, J_M$. However this implies that $c_{-n-s} \in J_1 \cup J_N$, a contradiction. Thus $f^{s+1}(J_M)$ is one of the subintervals $J_1, ..., J_N$ and $f^{s+1} : J_M \to f^{s+1}(J_M)$ is an order-preserving homeomorphism. If $f^{s+1}(J_M) = J_1$, then $f^{s+2}(J_1) \subset J_1$. So c^* must be a discontinuity point of f^{s+2}, otherwise it would be a periodic point of f. By Lemma 3.1 we have $c^* = c_{-m}$ for some integer $m \in [1, s+1]$. We also have $f^{s+2}(c^{**}) = 0$, and hence $c^{**} = c_{-s-1}$ and $m \leq s$. However the right endpoint of $f^m(J_M)$ is

$$\begin{aligned} f^{m+1}(c_{-m}-) &= \widetilde{f^{m+1}}(c_{-m}) \\ &= \tilde{f}^{m+1}(c_{-m}) = 1, \end{aligned}$$

contrary to $f^m(J_M) \neq J_N$. If $f^{s+1}(J_M) = J_N$, then clearly $c^{**} \in \mathrm{Per}(f)$, a contradiction. We thus conclude that (1) and (2) hold for $k = s+1$, which completes the

† The "invariance" means merely that the corresponding subinterval will never be divided by forthcoming inverse images of c. So this remains to be one of the subintervals for any level $n \geq N$.

induction.

It follows from (1) and (2) that there exist integers $r \geq 1$ and $\ell \in [2, N-1]$ satisfying $f^r(J_\ell) = J_\ell$. Hence the left endpoint of J_ℓ is a periodic point of f. This contradiction shows the following

Lemma 6.2. *For any $f \in \mathfrak{B}_\infty$ Case* (a) *occurs for infinitely many n's.*

6.4 Cases (A) and (B)

It follows from Lemmas 6.1 and 6.2 that there exist subsequences $\{p_n\}, \{q_n\}$ in $\mathbb{N}$ such that $q_1 < q_2 < q_3 < \cdots$, $1 \leq p_n < q_n$, $\gcd(p_n, q_n) = 1$ and that

$$\sum_{i=0}^{k-1} \epsilon_i(0) = \left[\frac{p_n}{q_n} k \right]$$

for any $1 \leq k < q_n$. In particular, since

$$\left[\frac{p_n}{q_n} k \right] = \left[\frac{p_{n+1}}{q_{n+1}} k \right]$$

holds for any $1 \leq k < q_n$, the fractions p_n/q_n and p_{n+1}/q_{n+1} belong to the same Farey interval of order $q_n - 1$ by Lemma 5.5 and we denote the corresponding Farey interval by L_n. Note that both p_n/q_n and p_{n+1}/q_{n+1} are interior points of L_n. Clearly $\overline{L_n}$ makes a descending sequence of closed intervals. Using Cantor's intersection theorem, we see that the limit set

$$\bigcap_{n=1}^{\infty} \overline{L_n}$$

is not empty and consists of a single point, say λ, because $|L_n| \to 0$ as $n \to \infty$.

Let ξ_n be the right endpoint of L_n, which makes a monotonically decreasing sequence converging to λ. We then distinguish two cases as follows:

(A) $\xi_n > \lambda$ for any n.

Since λ and p_n/q_n belong to the same Farey interval of order $q_n - 1$, we have $[p_n k/q_n] = [\lambda k]$ for $1 \leq k < q_n$. Therefore

$$\sum_{i=0}^{k-1} \epsilon_i(0) = [\lambda k] \tag{6.1}$$

for any $k \geq 1$, because n is arbitrary. This case occurs if $\lambda \notin \mathbb{Q}$. It may be possible for λ to be the left endpoint of L_n for all sufficiently large n. If so, λ is a rational number in the interval $[0, 1)$.

(B) $\xi_n = \lambda$ for some N.

By definition we have $\xi_n = \lambda$ for all $n \geq N$. In this case λ is a rational number in the interval $(0, 1]$, say r/s. Of course, we put $r = s = 1$ when $\lambda = 1$. Putting $L_n = [t/u, r/s) \in \mathcal{I}_{q_n-1}$ for $n \geq N$, we have

$$\frac{p_n}{q_n} \in \left[\frac{t}{u}, \frac{r}{s}\right)$$

and $q_n r - p_n s = 1$, because p_n/q_n must be the mediant of t/u and r/s in the Farey series $\mathfrak{F}_{q_n}$. Since

$$0 < \lambda k - \frac{p_n}{q_n}k = \frac{k}{q_n s} < \frac{1}{s}$$

for $1 \leq k < q_n$, we have $\lceil \lambda k \rceil = \lceil p_n k/q_n \rceil = [p_n k/q_n] + 1$. So, we get

$$\sum_{i=0}^{k-1} \epsilon_i(0) = \lceil \lambda k \rceil - 1 \tag{6.2}$$

for all $k \geq 1$, because $n \geq N$ is arbitrary.

The expressions (6.1) and (6.2) look somewhat different. However they are the unique integers in the interval $(\lambda k - 1, \lambda k]$ and $[\lambda k - 1, \lambda k)$ respectively; that is, the difference is merely at endpoints.

In both cases λ coincides with the rotation number $\rho(f)$ by Theorem 3.2. So, rounding up the above arguments we obtain the following

Lemma 6.3. *For any $f \in \mathfrak{B}_\infty$ we have*

$$\epsilon_n(0) = [\rho(n+1)] - [\rho n]$$

for all $n \geq 0$ in Case (A), *and*

$$\epsilon_n(0) = \lceil \rho(n+1) \rceil - \lceil \rho n \rceil$$

for all $n \geq 1$ in Case (B), *where $\rho = \rho(f)$ is the rotation number of f.*

6.5 Invariant Coordinates for $f \in \mathfrak{B}_\infty$

From Lemma 6.3 we have the following theorem. Compare with Theorem 4.3 for the case $f \in \mathfrak{E}(p/q)$.

Theorem 6.1. *For any $f \in \mathfrak{B}_\infty$ with $\rho(f) = p/q \in [0, 1] \cap \mathbb{Q}$ we have*

$$\Theta_f(0, t) = \frac{Q_{p,q}(t)}{1 - t^q}$$

where

$$Q_{p,q}(t) = \sum_{\substack{1 \le k < q \\ \{pk/q\} + p/q \ge 1}} t^k \tag{6.3}$$

in Case (A), *and*

$$\Theta_f(0,t) = \frac{Q^*_{p,q}(t)}{1 - t^q}$$

where

$$Q^*_{p,q}(t) = \begin{cases} t & (p/q = 1/1), \\ Q_{p,q}(t) + t^q - t^{q-1} & (0 < p/q < 1), \end{cases}$$

in Case (B).

Proof. In Case (A) p/q belongs to $[0, 1)$ and the proof is similar to that of Theorem 4.3 if $0 < p/q < 1$, because the sequence $\{\epsilon_n(0)\}_{n \ge 0}$ is periodic with period q. It is valid even when $p/q = 0/1$ because $Q_{0,1}(t) = 0$.

In Case (B) p/q belongs to $(0, 1]$. We have $Q^*_{1,1}(t) = t$ and

$$Q^*_{p,q}(t) = \sum_{\substack{1 \le k \le q \\ \lceil p(k+1)/q \rceil - \lceil pk/q \rceil = 1}} t^k,$$

if $0 < p/q < 1$. Since $\lceil p(k+1)/q \rceil - \lceil pk/q \rceil = 1$ occurs if and only if

$$\frac{p(k+1)}{q} \ge \left\lceil \frac{p}{q} k \right\rceil + \frac{1}{q}$$

and since $\lceil pk/q \rceil = [pk/q] + 1$ if $pk/q \notin \mathbb{Z}$, we have

$$Q^*_{p,q}(t) = t^q + \sum_{\substack{1 \le k < q \\ \{pk/q\} + p/q \ge 1 + 1/q}} t^k. \tag{6.4}$$

However $k = q - 1$ is counted in the summation in the right-hand side of (6.3) but not in that of (6.4), which is a unique integer in the interval $[1, q - 1]$ satisfying the equality $\{pk/q\} + p/q = 1$. This completes the proof. □

Note that $\Theta_f(0,t)$ has the following asymptotic development in Case (B):

$$\Theta_f(0,t) = \frac{p}{q} \cdot \frac{1}{1-t} - \frac{1}{2} - \frac{1}{2q} + O(t-1) \tag{6.5}$$

as $t \to 1$, because $Q^*_{p,q}(1) = p$ and

$$Q^{*\prime}_{p,q}(1) = Q'_{p,q}(1) + 1$$
$$= \frac{(p+1)(q-1)}{2} + 1.$$

For any irrational $\mu > 0$ we introduce the power series

$$H_\mu(z) = \sum_{n=1}^{\infty} [\mu n] z^n. \tag{6.6}$$

The radius of convergence of H_μ is clearly 1. The following is a straightforward consequence of Lemma 6.3.

Theorem 6.2. *For any $f \in \mathfrak{B}$ with $\rho = \rho(f) \notin \mathbb{Q}$ we have*

$$\Theta_f(0,t) = \frac{1-t}{t} H_\rho(t).$$

The following theorem shown by Hecke (1921) means that, if the rotation number of $f \in \mathfrak{B}$ is irrational, then the invariant coordinate $\Theta_f(0,t)$ is not a rational function. We give the proof for the sake of completeness.

Theorem 6.3. *For any irrational $\mu > 0$ the unit circle $|z| = 1$ is the natural boundary of the power series $H_\mu(z)$.*

Proof. We consider the power series $H^*_\mu(z)$ obtained by replacing $[\mu n]$ with $\{\mu n\}$ in (6.6), which differs from $H_\mu(z)$ only by a rational function. So, it suffices to show that

$$z_m = \exp(2\pi i \mu m)$$

is a singular point of H^*_μ for each integer m, because the sequence $\{z_m\}$ forms a dense subset on the unit circle. For any $0 < r < 1$ put

$$H^*_\mu(r z_m) = \sum_{n=1}^{\infty} \{\mu n\} r^n \exp(2\pi i \mu m n)$$
$$= \sum_{n=1}^{\infty} a_n r^n$$

and $s_n = a_1 + \cdots + a_n$ for $n \geq 1$. Since

$$s_n = \sum_{k=1}^{n} \{\mu k\} \exp(2\pi i \mu m k)$$
$$= \sum_{k=1}^{n} \{\mu k\} \exp(2\pi i m \{\mu k\}),$$

we have

$$s_n = \sum_{k=1}^{n} \varphi_m(\{\mu k\})$$

where

$$\varphi_m(x) = x \exp(2\pi i m x) \in C[0, 1].$$

Since the sequence $\{\mu k\}$ is distributed uniformly on the unit interval $[0, 1]$, we get

$$\lim_{n\to\infty} \frac{s_n}{n} = \int_0^1 \varphi_m(x)\,dx = \begin{cases} \dfrac{1}{2\pi i m} & (m \neq 0), \\ \dfrac{1}{2} & (m = 0). \end{cases}$$

Applying Lemma 3.3 to the power series

$$a_1 r + (a_2 - a_1) r^2 + (a_3 - a_2) r^3 + \cdots = (1 - r) \sum_{n=1}^{\infty} a_n r^n,$$

it follows that the limit

$$\lim_{r\to 1-} (1 - r) H_\mu^*(r z_m)$$

exists and does not vanish. Hence z_m is a singular point of H_μ^*, as required. □

In particular, putting $m = 0$, we have

$$\lim_{r\to 1-} (1 - r) H_\mu^*(r) = \frac{1}{2}$$

for any irrational μ, which implies that

$$\Theta_f(0, t) = -\frac{\rho(f)}{t - 1} - \frac{1}{2} + o(1) \tag{6.7}$$

as $t \to 1-$ for any f with irrational $\rho(f)$.

Putting together the formulae (4.2), (6.5) and (6.7), we conclude that the limit

$$\chi_f = \lim_{t\to 1-} \left(\Theta_f(0, t) - \frac{\rho(f)}{1 - t} + \frac{1}{2} \right)$$

exists for any $f \in \mathfrak{B}$, where

$$\chi_f = \begin{cases} \pm\dfrac{1}{2q} & \left(\rho(f) = \dfrac{p}{q}\right), \\ 0 & (\rho(f) \notin \mathbb{Q}). \end{cases}$$

It is an easy exercise to show that $|\chi_f|$, as a function of $\rho(f)$, is continuous at every irrational points and discontinuous at every rational points.

Exercises in Chapter 6

1 Give an example of $f \in \mathfrak{B}_\infty$ satisfying

$$\cdots < c_{-2n} < \cdots < c_{-4} < c_{-2} < c_0 < \cdots < c_{-2n-1} < \cdots < c_{-3} < c_{-1},$$

where c_{-2n} and c_{-2n-1} converge to $\alpha \in (0, c)$ and $\beta \in (c, 1)$ respectively. Then observe that f satisfies Case (A) with

$$\xi_n = \frac{n}{2n-1} > \frac{1}{2} = \rho(f) \quad \text{and} \quad \Theta_f(0, t) = \frac{t}{1-t^2}.$$

2 Give an example of $f \in \mathfrak{B}_\infty$ satisfying

$$c_{-1} < c_{-3} < \cdots < c_{-2n-1} < \cdots < c_0 < c_{-2} < c_{-4} < \cdots < c_{-2n} < \cdots,$$

where c_{-2n-1} and c_{-2n} converge to $\alpha \in (0, c)$ and $\beta \in (c, 1)$ respectively. Then observe that f satisfies Case (B) with

$$\xi_n = \frac{1}{2} = \rho(f) \quad \text{and} \quad \Theta_f(0, t) = \frac{t^2}{1-t^2}.$$

3 For any real number λ in the interval $[0, 1]$ we define

$$\mathfrak{S}^+(\lambda) = (a_1^+, a_2^+, a_3^+, \ldots) \in \{0, 1\}^{\mathbb{N}}$$

where $a_n^+ = \lceil \lambda(n+1) \rceil - \lceil \lambda n \rceil$. Show that $\mathfrak{S}^+(p/q)$ coincides with the tail inversion sequence $\mathfrak{S}^*(p/q)$ for any irreducible fraction p/q in the interval $(0, 1)$.

4 The Thue-Morse sequence

$$\{t_n\}_{n \ge 0} = 01101001100101101001011001101001\ldots$$

is defined by $1 - 2t_n = (-1)^{\sigma_n}$, where σ_n is the sum of digits in the binary expansion of n; equivalently,

$$\sum_{n=0}^{\infty} (1 - 2t_n) x^n = \prod_{n=0}^{\infty} (1 - x^{2^n}).$$

Show that there is no $f \in \mathfrak{B}$ satisfying $\epsilon_n(0) = t_n$ for all $n \ge 0$.

Chapter 7

Limit Sets Ω_f and $\omega_f(x)$

Two kinds of limit sets for f in $\mathfrak{B}_\infty$ are introduced and investigated; one concerns the future of images $f^n(J)$, the other concerns the future of an orbit $f^n(x)$. It is shown that every itinerary $\eta(x)$ is not periodic if the rotation number of f is irrational.

7.1 Definition of Ω_f

For $f \in \mathfrak{B}_\infty$ we evidently have a descending sequence of compact sets

$$[0,1] \supset \overline{f(J)} \supset \overline{f^2(J)} \supset \overline{f^3(J)} \supset \cdots,$$

where each $\overline{f^n(J)}$ is the union of $a_n(f) = n + 1$ disjoint closed intervals. We then define the *limit set*

$$\Omega_f = \bigcap_{n=0}^{\infty} \overline{f^n(J)},$$

being a non-empty compact set by Cantor's intersection theorem. Since

$$\overline{f^n(J)} = [0,1] \setminus \bigcup_{k=1}^{n} \operatorname{Int} K_k$$

by (iii) of Lemma 4.3, it follows that

$$\Omega_f = [0,1] \setminus \bigcup_{n=1}^{\infty} \operatorname{Int} K_n. \tag{7.1}$$

This means that the set Ω_f is the remainder obtained from the interval $[0, 1]$ by an infinite sequence of deletions of open intervals, suggestive of the classical Cantor's ternary set. However, the limit set Ω_f is not necessarily totally disconnected. For example, the limit set of the function f in Example 3.1 contains the interval $[p_1, p_2]$, because $[p_1, p_2] = f^n([p_1, p_2]) \subset \overline{f^n(J)}$ for all $n \geq 0$.

Note that both $\mathrm{Orb}_f(0)$ and $\mathrm{Orb}_f(1)$ are contained in Ω_f, as well as all inverse images c_{-n} by (ii) of Lemma 4.3 for any $f \in \mathfrak{B}_\infty$. The image $f(\Omega_f)$ is a proper subset of Ω_f, because Ω_f contains 1 while $f(\Omega_f)$ does not. We also have

$$|\Omega_f| = \lim_{n\to\infty} |f^n(J)| = 1 - \sum_{n=1}^{\infty} |K_n|. \tag{7.2}$$

As a consequence of the expression (7.1), we have the following

Lemma 7.1. *The limit set Ω_f is perfect.*

Proof. It suffices to show that Ω_f has no isolated points. Suppose, on the contrary, that x is an isolated point of Ω_f. If $x \in (0, 1]$, then there exists a non-empty open interval $U = (x', x)$ with $U \cap \Omega_f = \emptyset$. Hence there exists an integer $i \geq 1$ satisfying $U \cap \mathrm{Int}\, K_i \neq \emptyset$ from (7.1). Since both $f^i(0)$ and x belong to Ω_f, we must have $x = f^i(0)$ and so $x < 1$. Applying the same argument to the right-hand side of x if $x \in [0, 1)$, we get $x = f^j(1)$ and so $x > 0$. Therefore $x \in (0, 1)$, $i \neq j$ and $f^i(0) = f^j(1)$, contrary to the fact that $\overline{K_i} \cap \overline{K_j} = \emptyset$ by (i) of Lemma 4.3. □

We next give a simple lemma. Let κ_n be the maximal length of $n+1$ intervals constituting $f^n(J)$. It is obvious that $\{\kappa_n\}$ is a monotonically decreasing sequence.

Lemma 7.2. *The limit set Ω_f is totally disconnected if and only if $\kappa_n \to 0$ as $n \to \infty$.*

Proof. If Ω_f has a connected component of length $\delta > 0$, it is clear that $\kappa_n \geq \delta$. This contraposition implies that Ω_f is totally disconnected if $\kappa_n \to 0$ as $n \to \infty$. Conversely, suppose that Ω_f is totally disconnected. Let W be an interval in $f^n(J)$ of length κ_n. We divide W into three parts equally, say W_1, W_2 and W_3. Since Ω_f is totally disconnected, there exist some $n_1, n_2, n_3 > n$ with $W_i \cap K_{n_i} \neq \emptyset$ for $1 \leq i \leq 3$. Applying this argument to other components of $f^n(J)$, we get $\kappa_N/\kappa_n < 2/3$ for some $N > n$, and hence κ_n must converge to 0 as $n \to \infty$. □

Theorem 7.1. *If Ω_f is totally disconnected, then it is homeomorphic to Cantor's ternary set.*

Proof. A homeomorphism h from Cantor's ternary set onto Ω_f will be constructed as follows: Put $h(1/3) = f(1)$, $h(2/3) = f(0)$ and define h on the first middle third $[1/3, 2/3]$ linearly so that h maps $[1/3, 2/3]$ onto K_1 homeomorphically. Next let K_i, $i \geq 2$ be the first one among K_n's belonging to the left component of $[0, 1] \setminus \mathrm{Int}\, K_1$. We then put $h(1/9) = f^i(1)$, $h(2/9) = f^i(0)$ and define h on $[1/9, 2/9]$ linearly. Similarly we define h on $[7/9, 8/9]$ linearly so that $h([7/9, 8/9]) = K_j$, where K_j is the first one belonging to the right component of

$[0, 1] \setminus \text{Int}\, K_1$. The function h is thus defined on all middle thirds by repeating this process infinitely many times. Moreover h can be extended continuously to $[0, 1]$, because Ω_f is perfect and totally disconnected. □

For simplicity we put

$$\Omega_f^\circ = \Omega_f \setminus (\text{Orb}_f(0) \cup \text{Orb}_f(1)),$$

which is neither countable nor compact. By Corollary 4.1 it is easily verified that

$$\Omega_f^\circ = \bigcap_{n=0}^{\infty} \text{Int}\, f^n(J).$$

As a straightforward corollary of Theorem 7.1, we have

Corollary 7.1. *If Ω_f is totally disconnected, then each $x \in \Omega_f$ is an accumulation point of $\{K_n\}$. Moreover, for any $x \in \Omega_f^\circ$ we can take a subsequence of $\{K_n\}$ converging to x on each side of x.*

7.2 Aperiodic Behavior

We first need the following

Lemma 7.3. *The following three statements are equivalent with respect to one another for $f \in \mathfrak{B}_\infty$:*

(i) $\rho(f) \notin \mathbb{Q}$.
(ii) $\text{Per}(f) = \emptyset$.
(iii) $e_n > 0$ *for all* $n \geq 1$, *where*

$$e_n = \inf_{0 \leq x < 1} |x - f^n(x)|;$$

in other words, the graph of f^n separates from the diagonal.

Proof. Obviously $\text{Per}_q(f) \neq \emptyset$ for some $q \geq 1$ implies $\rho(f) \in \mathbb{Q}$ and $e_q = 0$. This contraposition implies that (i), as well as (iii), derives (ii).

Suppose next that $\text{Per}(f) = \emptyset$. If $\rho(f) \in \mathbb{Q}$, then $f^s(1) = 0$ holds for some $s \geq 1$ by Theorem 3.4, and hence f belongs to the subclass $\mathfrak{E}^\phi$ by Theorem 4.7. This contradiction implies (i). On the other hand, suppose that $e_q = 0$ for some $q \geq 1$, where q is assumed to be the smallest integer with this property. Then there exists some discontinuity point $z \in (0, 1)$ of f^q satisfying $f^q(z-) = z$, and thus $z \in \text{Per}_q(\tilde{f})$. From Lemma 3.1 we have $z = c_{-k}$ for some integer $k \in [0, q)$. Since

$f(c_{-m}) = \tilde{f}(c_{-m})$ for $m \geq 1$, it follows that

$$\begin{aligned}\tilde{f}^{k+1}(z) &= \tilde{f}^{k+1}(c_{-k}) \\ &= \tilde{f} \circ f^k(c_{-k}) \\ &= \tilde{f}(c) = 1,\end{aligned}$$

which implies that $1 \in \mathrm{Per}_q(\tilde{f})$ and $\tilde{f}^{q-1}(1) = c$. Thus we have $c_{1-q} = 1$, because

$$\tilde{f}^{q-1}(c_{1-q}) = f^{q-1}(c_{1-q}) = c$$

and $\tilde{f}$ is a one-to-one mapping defined on $(0, 1]$. Hence

$$f^q(1) = f^q(c_{1-q}) = 0$$

and f belongs to the subclass $\mathfrak{C}^{\phi}$. This contradiction implies (iii). □

Theorem 7.2. *If Ω_f is totally disconnected, then the rotation number $\rho(f)$ is irrational.*

Proof. It suffices to show that f has no periodic points by (i) and (ii) of Lemma 7.2. Suppose, on the contrary, that $z \in [0, 1)$ is a periodic point with period $d \geq 1$. Since $f^d(x)$ is continuous at z, there exists a sufficiently small open neighborhood U of z such that $U, \ldots, f^{d-1}(U)$ are disjoint and that $f^k : U \to f^k(U)$ is an order-preserving homeomorphism for each $1 \leq k \leq d$. Since $z \in \Omega_f^{\circ}$, there exist two subsequences of K_n approaching z from above and below respectively. Take now some K_m in U on the right side of z. We distinguish two cases as follows.

(a) $z < f^{m+d}(1) < f^m(1)$.

Let z' be the largest fixed point of f^d in the interval $[z, f^m(1)]$. The graph of $f^d(x)$ is below the diagonal on $(z', f^m(1)]$. Since f^d is strictly increasing, we have an infinite decreasing sequence of intervals satisfying

$$z \prec \cdots \prec K_{m+2d} \prec K_{m+d} \prec K_m.$$

Then f^k maps these intervals to

$$f^k(z) \prec \cdots \prec K_{m+k+2d} \prec K_{m+k+d} \prec K_{m+k}$$

in $f^k(U)$ for each $1 \leq k < d$. This clearly contradicts that there is a subsequence of K_n converging to z on the left side of z.

(b) $f^m(1) < f^{m+d}(1)$.

In this case the graph of $f^d(x)$ is above the diagonal on $(z', f^m(1)]$. Therefore there exists an inverse image of $f^m(1)$ by f^{kd} for all $k \geq 1$. In particular, we have $f^{-i}(1)$ for some $i \geq 1$, contrary to the fact that 1 has no inverse images.

We arrive in the contradiction in either case, which completes the proof. □

Theorem 7.3. *If $\rho(f) \notin \mathbb{Q}$, then the itinerary $\eta(x)$ is not periodic for any $x \in [0, 1)$.*

Proof. Suppose, on the contrary, that there exists some $z \in [0, 1)$ such that $\eta(z)$ is a periodic itinerary with period $d \geq 1$. For $0 \leq k < d$ we put

$$\alpha_k = \inf_{n \geq 0} f^{dn+k}(z) \quad \text{and} \quad \beta_k = \sup_{n \geq 0} f^{dn+k}(z).$$

If $\alpha_k = \beta_k$, then z would be a periodic point of f, contrary to (ii) of Lemma 7.3, and hence each interval $L_k = [\alpha_k, \beta_k)$ is not empty. Since $\eta(z)$ is periodic with period d, we have $c \notin [\alpha_k, \beta_k)$ for every k. Therefore each restriction $f|L_k$ is an order-preserving homeomorphism, and so,

$$f(L_0) = L_1, \ldots, f(L_{d-2}) = L_{d-1} \quad \text{and} \quad f(L_{d-1}) \subset L_0.$$

However $f^d(L_0) \subset L_0$ implies that the graph of f^d touches the diagonal, contrary to (iii) of Lemma 7.3. □

A typical example of $f \in \mathfrak{B}_\infty$ having fixed or periodic points is illustrated in Fig. 3.1.

7.3 ω-Limit Sets

For each $x \in [0, 1]$ we have a descending sequence of compact sets

$$\overline{\mathrm{Orb}_f(x)} \supset \overline{\mathrm{Orb}_f(f(x))} \supset \overline{\mathrm{Orb}_f(f^2(x))} \supset \cdots.$$

The *ω-limit set* of x is defined by

$$\omega_f(x) = \bigcap_{n=0}^{\infty} \overline{\mathrm{Orb}_f(f^n(x))} = \bigcap_{n=0}^{\infty} \overline{\bigcup_{m=n}^{\infty} \{f^m(x)\}},$$

being a non-empty compact set. In other words, $y \in \omega_f(x)$ if and only if there exists a subsequence $\{n_j\}$ such that $f^{n_j}(x)$ converges to y as $j \to \infty$. In particular, the set of all accumulation points of $\mathrm{Orb}_f(x)$ is contained in $\omega_f(x)$.

Since $\mathrm{Orb}_f(f^n(x)) \subset f^n(J)$, we have $\omega_f(x) \subset \Omega_f$ for any $x \in [0, 1]$. To study this limit set we need the following lemma. Compare with Corollary 7.1.

Lemma 7.4. *Suppose that Ω_f is totally disconnected. Then every $x \in \Omega_f$ is an accumulation point of $\{c_{-n}\}$. Moreover, for $x \in \Omega_f^\circ$ we can take a subsequence of $\{c_{-n}\}$ converging to x on each side of x.*

Proof. By Corollary 7.1 it suffices to show that any non-empty interval in the form $E = [f^n(0), f^m(1))$ with $n \neq m$ contains at least one inverse image of c. Suppose, on the contrary, that E contains no inverse images of c. Then each restriction $f^k | E$ is an order-preserving homeomorphism onto its image. Recall that every interval $\overline{K_i}$ also contains no inverse images of c. If $n < m$, then putting $d = m - n$, we see that

$$W = \bigcup_{k=0}^{\infty} K_{n+kd} \cup \bigcup_{k=0}^{\infty} f^{kd}(E)$$

is a right-open interval containing no inverse images of c satisfying $f^d(W) \subset W$. This implies that the graph of f^d touches the diagonal, contrary to (iii) of Lemma 7.3. We can treat the case in which $n > m$ similarly. This completes the proof. □

By Corollary 4.1 and Lemma 7.4 we immediately have

Corollary 7.2. *Each interval constituting $f^n(J)$ contains at least one inverse image of c, if Ω_f is totally disconnected.*

Using the above lemma we have the following main result in the section.

Theorem 7.4. *If Ω_f is totally disconnected, then $\omega_f(x) = \Omega_f$ for any $x \in [0, 1]$.*

Proof. We first show that $\omega_f(x) = \omega_f(0)$ for any

$$x \in \{0\} \cup \{1\} \cup \bigcup_{n=1}^{\infty} \overline{K}_n.$$

To see this, let $x \in \overline{K_n}$ for $n \geq 1$. As is discussed in Section 6.1, $\overline{K_n}$ contains no inverse images of c and $f^k : \overline{K_n} \to \overline{K_{n+k}}$ is a homeomorphism for all $k \geq 1$. Hence

$$\lim_{k\to\infty} \left| f^k(x) - f^{n+k}(0) \right| \leq \lim_{k\to\infty} |K_{n+k}| = 0$$

by (7.2), from which we have $\omega_f(x) = \omega_f(0)$. Similarly we get $\omega_f(1) = \omega_f(0)$, because

$$\lim_{k\to\infty} \left| f^k(0) - f^k(1) \right| = \lim_{k\to\infty} |K_k| = 0.$$

We next show that $\omega_f(x) = \Omega_f$ for any $x \in \Omega_f^\circ$. Let $m \geq 1$ be an arbitrarily fixed integer. Since $f^m(x) \in \Omega_f^\circ$, there exists a subsequence $\{n_j\}$ such that c_{-n_j} converges to $f^m(x)$ on the right side as $j \to \infty$ by Lemma 7.4. Since $\kappa_n \to 0$ as $n \to \infty$ by Lemma 7.2, it follows from Corollary 7.2 that there exists an integer

$m' > m$ such that $\kappa_\ell < \kappa^*$ for any $\ell \geq m'$, where

$$\kappa^* = \min_{1 \leq k \leq m} |K_k|,$$

and that each interval constituting $f^m(J)$ contains at least one of $c_0, \ldots, c_{-m'}$. Take a sufficiently large integer $N > m'$ so that

(1) $c_{-i} \notin \operatorname{Int} V$ for any $0 \leq i < N$, where $V = [f^m(x), c_{-N}]$,

(2) $|f^j(V)| < \kappa^*$ for any $0 \leq j < m'$.

Such an integer N certainly exists, because each f^j is right-continuous. Then, by (1), each restriction $f^j | V$ is an order-preserving homeomorphism onto its image for $1 \leq j \leq N$. In particular, each image $f^j(V)$ is a closed interval. By (2) we have

$$|f^{j+m}(x) - f^j(c_{-N})| = |f^j(V)| < \kappa^*$$

for $0 \leq j < m'$. Moreover, for $m' \leq j \leq N$ we have

$$|f^{j+m}(x) - f^j(c_{-N})| = |f^j(V)| \leq \kappa_j < \kappa^*,$$

because $f^j(V)$ is contained in some connected component of $f^j(J)$. The above inequalities show that $f^{j+m}(x)$ and c_{j-N} belong to the same interval constituting $f^m(J)$ for all $0 \leq j \leq N$. Since every interval constituting $f^m(J)$ contains at least one of $c_0, \ldots, c_{-m'}$, it also contains at least one of $f^{m+N-m'}(x), \ldots, f^{m+N}(x)$. Since m is arbitrary and $\operatorname{Per}(f) = \emptyset$ by Lemma 7.3 and Theorem 7.2, we conclude that every point in Ω_f is an accumulation point of $\operatorname{Orb}_f(x)$; that is, $\omega_f(x) = \Omega_f$, as required.

It follows from Theorem 7.1 that there exists a subsequence of $\{K_n\}$ converging to $f^m(0)$ on the right side for any $m \geq 1$. Applying the above argument to $f^m(0)$ instead of $f^m(x)$, we get $\omega_f(0) = \Omega_f$. This completes the proof. □

7.4 Recurrence

A point $x \in [0, 1]$ is said to be *non-wandering* for f provided that, for any neighborhood U of x, there exists $y \in U$ and $n \geq 1$ satisfying $f^n(y) \in U$. On the other hand, a point x is said to be *recurrent* for f provided that, for any neighborhood U of x, there exists $n \geq 1$ satisfying $f^n(x) \in U$. For example, periodic points are thus non-wandering and recurrent.

We now identify non-wandering and recurrent points for $f \in \mathfrak{B}_\infty$.

Theorem 7.5. *Suppose that Ω_f is totally disconnected. Then Ω_f coincides with* *the set of all non-wandering points and also with that of all recurrent points for f.*

Proof. Let $x \in \Omega_f$. Since $\omega_f(x) = \Omega_f$ by Theorem 7.4, there exists a subsequence $\{n_j\}$ such that $f^{n_j}(x)$ converges to x as $j \to \infty$. For any neighborhood U of x, we can find some n_j and n_{j+1} satisfying

$$y = f^{n_j}(x) \in U \quad \text{and} \quad f^{n_{j+1}-n_j}(y) = f^{n_{j+1}}(x) \in U,$$

and hence the point x is both non-wandering and recurrent for f.

On the other hand, if $x \notin \Omega_f$, then $x \in \operatorname{Int} K_n$ for some $n \geq 1$. So we can take a neighborhood U of x with $U \subset K_n$. Since $f^m(U) \subset K_{m+n}$ for any $m \geq 1$, we have

$$U \cap f^m(U) = \emptyset,$$

and thus x is neither non-wandering nor recurrent. This completes the proof. □

7.5 Complexity of Itineraries

As in the case with orbits, we introduce the equivalence relation in the sense of Kuratowski in the set of itineraries

$$\Sigma_f = \{\eta(x) : 0 \leq x \leq 1\}$$

for $f \in \mathfrak{B}_\infty$. We define $\eta(x) \sim \eta(y)$ if there exist non-negative integers m and n satisfying

$$\eta \circ f^m(x) = \eta \circ f^n(y).$$

In other words, if $\eta(x)$ coincides with $\eta(y)$ except for their first finite terms, then they are equivalent to one another. We show that there are so many equivalence classes for some $f \in \mathfrak{B}_\infty$.

Theorem 7.6. *Suppose that Ω_f is totally disconnected. Then there are uncountable equivalence classes in the sense of Kuratowski in Σ_f.*

Proof. Let $x \neq y$ be any two points in the uncountable set

$$\Omega_f^\circ = \Omega_f \setminus (\operatorname{Orb}_f(0) \cup \operatorname{Orb}_f(1))$$

with $\eta(x) \sim \eta(y)$. By definition there exist non-negative integers m, n satisfying

$$\eta \circ f^m(x) = \eta \circ f^n(y).$$

Putting $x' = f^m(x)$ and $y' = f^n(y)$, we have $\eta(x') = \eta(y')$. If $x' = y'$, then

$$\mathrm{Orb}_f(x) \cap \mathrm{Orb}_f(y) \neq \emptyset,$$

and hence $x \sim y$ as orbits. Otherwise, the open interval (x', y') contains no inverse images of c, because $\eta(x') = \eta(y')$. By Lemma 7.4 we have $x', y' \notin \Omega_f^\circ$. Since $f(\Omega_f) \subset \Omega_f$ (see Exercise 2 in this chapter), it follows that $x', y' \in \Omega_f$ and hence

$$x', y' \in \mathrm{Orb}_f(0) \cup \mathrm{Orb}_f(1).$$

Suppose now that $x' \in \mathrm{Orb}_f(1)$. There exists some integer $s \geq 0$ satisfying

$$x' = f^m(x) = f^s(1).$$

Since $f([0,1]) \subset [0,1)$, we must have $m \leq s$ and so

$$x = f^{s-m}(1) \in \mathrm{Orb}_f(1),$$

contrary to $x \in \Omega_f^\circ$. Thus $x' \notin \mathrm{Orb}_f(1)$ and the same argument as above implies that $y' \notin \mathrm{Orb}_f(1)$. Hence both x' and y' belong to $\mathrm{Orb}_f(0)$, which means that $x \sim y$ as orbits.

Taking the contraposition, we see that $x \nsim y \in \Omega_f^\circ$ implies $\eta(x) \nsim \eta(y)$. Since there are uncountable different orbits in the sense of Kuratowski by Lemma 1.1, the set of itineraries Σ_f possesses the same property. □

Exercises in Chapter 7

1 Prove that Cantor's ternary set is the set of all real numbers $x \in [0, 1]$, which can be expressed as

$$x = \sum_{n=1}^{\infty} \frac{a_n}{3^n} = 0.a_1a_2a_3 \cdots (3)$$

where each a_n is either 0 or 2. For example, $1/3 = 0.1(3)$ belongs to Cantor's ternary set, because $1/3$ can also be expressed as $1/3 = 0.0222\cdots(3)$. [It remains an open question whether Cantor's ternary set contains a quadratic irrational or not.]

2 Show that $f(\Omega_f) \subset \Omega_f$ for any $f \in \mathfrak{B}_\infty$.

3 Show that $\eta(0) \sim \eta(1)$ in the sense of Kuratowski for any $f \in \mathfrak{B}$.

4 A non-empty subset X of the interval $[0, 1]$ is said to be an *invariant* set of $f : [0, 1] \to [0, 1]$ provided that $f(X) = X$. Show that

$$\bigcap_{n=0}^{\infty} f^n([0, 1))$$

is the maximal invariant set of $f \in \mathfrak{B}_\infty$.

5 Let X be the maximal invariant set of $f \in \mathfrak{B}_\infty$ stated in the previous exercise. Since $f(X) = X$, there exists a unique inverse image $f^{-n}(x)$ for any $x \in X$ and $n \geq 1$. Show that the backward average

$$\lim_{n\to\infty} \frac{1}{n} \sum_{k=0}^{n-1} \epsilon \circ f^{-k}(x)$$

exists and equals to $\rho(f)$ for any $x \in X$.

Chapter 8

Piecewise Linear Maps

For each piecewise linear map $\psi_{a,b,c}$ in $\mathfrak{E}^{\phi}(p/q)$ or $\mathfrak{E}^{0}(p/q)$ we give two basic formulae of the first kind. For each $\psi_{a,b,c}$ in $\mathfrak{B}_{\infty}$ we also give two basic formulae of the second kind, which have relation to the Hecke-Mahler series. We further investigate the Hausdorff dimension of the limit set for $\psi_{a,b,c}$ belonging to $\mathfrak{B}_{\infty}$.

8.1 Basic Formulae of the First Kind

In subsequent chapters we restrict ourselves to the piecewise linear map $\psi_{a,b,c}(x)$ defined in (2.5), which is also written as

$$\psi_{a,b,c}(x) = \left\{ \frac{a+b}{2}(x-c) - \frac{a-b}{2}\,|x-c| \right\}.$$

Then it is easily seen that the set of parameters (a, b, c) satisfying $\psi_{a,b,c} \in \mathfrak{B}(c)$ is

$$\Lambda = \left\{ (a,b,c) \in \mathbb{R}^3 : a, b > 0,\ ac + b(1-c) < 1,\ 0 < c < 1 \right\}.$$

Lemma 8.1. *For any $\psi_{a,b,c} \in \mathfrak{B}(c)$ we have $a^{1-\rho} b^{\rho} < 1$, where $\rho = \rho(\psi_{a,b,c})$.*

Proof. We simply write $\psi_{a,b,c}$ as ψ. We first assume that $\psi \in \mathfrak{E}(p/q)$ for some irreducible fraction $p/q \in (0, 1)$. As is shown in the proof of Theorem 4.2, each restriction $\psi^q \,|\, J_k$ is a linear function with slope $a^{q-p} b^p$. Since ψ^q is one-to-one, it follows that

$$|\psi^q(J)| = a^{q-p} b^p \le |\psi(J)| < 1.$$

This means that $a^{1-\rho} b^{\rho} < 1$ because of $\rho = \rho(\psi) = p/q$.

We next assume that $\psi \in \mathfrak{B}_{\infty}$. In the argument in Section 6.4 we have already seen that

$$\rho(n+1) - 1 \le \nu_n = \sum_{i=1}^{n} \epsilon \circ \psi^i(0) \le \rho(n+1)$$

for all $n \geq 1$ in both Cases (A) and (B). Put $K_k = [\psi^k(1), \psi^k(0))$ for any $k \geq 1$. The restriction $\psi^n | \overline{K_1}$ is a linear function with slope $a^{n-\nu_n} b^{\nu_n}$, because ν_n is the number of $i \in [1, n]$ satisfying $\psi^i(0) \geq c$. Therefore

$$\left| \psi^n(\overline{K_1}) \right| = |K_{n+1}| = a^{n-\nu_n} b^{\nu_n} |K_1|,$$

and hence†

$$\frac{|K_{n+1}|}{|K_1|} \geq \min\left(1, \frac{1}{a}, \frac{1}{b}\right) (a^{1-\rho} b^{\rho})^{n+1}.$$

Since $|K_n| \to 0$ as $n \to \infty$, we get $a^{1-\rho} b^{\rho} < 1$, as required. □

For any positive integers p, q we define the associated polynomials $P_{p,q}(w, z)$ and $P^+_{p,q}(w, z)$ by

$$P_{p,q}(w, z) = \sum_{k=1}^{q} w^{k-1} z^{[pk/q]}$$

and

$$P^+_{p,q}(w, z) = \sum_{k=1}^{q} w^{k-1} z^{\lceil pk/q \rceil - 1},$$

where $\lceil x \rceil$ is the ceiling function defined in Section 5.4. We will admit the case $p/q = 0/1$ exceptionally in Section 10.5.

The following formula is called the basic formula of the first kind, which represents the relation between the parameters a, b, c and the rotation number p/q for $\psi_{a,b,c} \in \mathfrak{C}^\phi(p/q)$. Note that c is uniquely determined for given $a \neq b$ and p/q.

Theorem 8.1. *For any $\psi_{a,b,c} \in \mathfrak{C}^\phi(p/q)$ we have*

$$P_{p,q}\left(a, \frac{b}{a}\right) = \frac{1 - a^{q-p} b^p}{1 - b - (a - b)c}.$$

Proof. We simply write $\psi_{a,b,c}$ as ψ. If $\mathrm{Per}_q(\psi) \neq \emptyset$, then the restriction $\psi^q | J_q$ would have a fixed point. This is a contradiction, because $\psi^{q-1}(1) = c$ and so $\widetilde{\psi^q}(1) = 1$. We thus have $\mathrm{Per}_q(\psi) = \emptyset$ and it follows from Theorem 4.5 that

$$\sum_{n=1}^{\infty} |K_n| = 1.$$

Since

$$|K_{n+1}| = a^{n-\nu_n} b^{\nu_n} |K_1|$$

† If $0 < a \leq 1$, then $a^{n-\nu_n} \geq a^{(n+1)(1-\rho)}$; otherwise, $a^{n-\nu_n} \geq a^{(n+1)(1-\rho)-1}$. Note that the case in which $a \geq 1$ and $b \geq 1$ does not occur.

with $\nu_n = [p(n+1)/q]$, we get

$$\frac{1}{|K_1|} = 1 + \sum_{n=1}^{\infty} a^{n-\nu_n} b^{\nu_n}$$
$$= -\frac{1}{a} + \sum_{n=0}^{\infty} a^{n-1-[pn/q]} b^{[pn/q]}.$$

Putting $n = sq + r$, $s \ge 0$, $0 \le r < q$, we have

$$\frac{1}{|K_1|} = -\frac{1}{a} + \sum_{s=0}^{\infty} (a^{q-p} b^p)^s \sum_{r=0}^{q-1} a^{r-1-[pr/q]} b^{[pr/q]},$$

which is equal to

$$-\frac{1}{a} + \frac{1}{1 - a^{q-p} b^p} \left(P_{p,q}\left(a, \frac{b}{a}\right) + \frac{1}{a} - a^{q-p-1} b^p \right) = \frac{P_{p,q}(a, b/a)}{1 - a^{q-p} b^p},$$

because $a^{q-p} b^p < 1$ by Lemma 8.1. Since $|K_1| = 1 - b - (a-b)c$, this completes the proof. □

We also have another basic formula of the first kind by taking account of only the intervals K_n lying in $[c, 1)$, which enables us to determine c even if $a = b$.

Theorem 8.2. *For any $\psi_{a,b,c} \in \mathfrak{E}^{\phi}(p/q)$ we have*

$$P^{+}_{q,p}\left(\frac{b}{a}, a\right) = \frac{a(1-c)(1 - a^{q-p} b^p)}{1 - b - (a-b)c}.$$

Proof. Noticing that $K_n \subset [c, 1)$ if and only if $\psi^n(0) > c$, we have

$$1 - c = \sum_{\substack{n \ge 1 \\ \psi^n(0) > c}} |K_n|$$
$$= \sum_{n=1}^{\infty} \epsilon \circ \psi^n(0) |K_n|$$
$$= |K_1| \sum_{n=1}^{\infty} \left(\left[\frac{p}{q}(n+1)\right] - \left[\frac{p}{q} n\right] \right) a^{n-1-[pn/q]} b^{[pn/q]}.$$

In the sum of the right-hand side we can add the term corresponding to $n = 0$ because it vanishes. Therefore, putting $n = sq + r$, $s \ge 0$, $0 \le r < q$, we have

$$\frac{1-c}{|K_1|} = \sum_{s=0}^{\infty} (a^{q-p} b^p)^s \sum_{r=0}^{q-1} \left(\left[\frac{p}{q}(r+1)\right] - \left[\frac{p}{q} r\right] \right) a^{r-1-[pr/q]} b^{[pr/q]}.$$

Since $0 < p/q < 1$, $[pr/q]$ takes all integers from 0 to $p - 1$ as $r \in [0, q)$ varies. Putting $[p(r+1)/q] = \ell$ and $[pr/q] = \ell - 1$, we have $pr/q < \ell \le p(r+1)/q$, and so $r < q\ell/p \le r+1$. Hence $r = \lceil q\ell/p \rceil - 1$ and

$$\frac{(1-c)(1-a^{q-p}b^p)}{|K_1|} = \sum_{\ell=1}^{p} a^{\lceil q\ell/p\rceil - \ell - 1} b^{\ell-1} = \frac{1}{a} P^+_{q,p}\left(\frac{b}{a}, a\right),$$

which completes the proof. □

Note that the basic formulae in Theorems 8.1 and 8.2 reduce to trivial identities when $a = b$ and $a = 1$ respectively.

There exist also two basic formulae of the first kind for $\psi_{a,b,c} \in \mathfrak{E}^0(p/q)$.

Theorem 8.3. *For any* $\psi_{a,b,c} \in \mathfrak{E}^0(p/q)$ *we have*

$$P^+_{p,q}\left(a, \frac{b}{a}\right) = \frac{1 - a^{q-p}b^p}{1 - b - (a-b)c}.$$

Proof. Since the piecewise linear function $\psi = \psi_{a,b,c}$ has at most one periodic orbit, it follows from Theorem 4.8 that

$$\sum_{n=1}^{\infty} |K_n^*| = 1.$$

Moreover c is the right endpoint of K_{q-1}^* and the left endpoint of K_n^* is $\psi^n(1)$ for all $n \ge 1$. Therefore

$$|K_{n+1}^*| = a^{n-\mu_n} b^{\mu_n} |K_1^*|$$

where $\mu_n = \epsilon \circ \psi(1) + \cdots + \epsilon \circ \psi^n(1)$. By (ii) of Theorem 4.6, the itinerary of $\psi(1)$ is the tail inversion sequence of $\mathfrak{S}(p/q)$. Thus, using Exercise 3 in Chapter 6, we have

$$\mu_n = \left\lceil \frac{p}{q}(n+1) \right\rceil - 1,$$

and therefore

$$\frac{1}{|K_1^*|} = 1 + \sum_{n=1}^{\infty} a^{n-\mu_n} b^{\mu_n} = -\frac{1}{b} + \sum_{n=0}^{\infty} a^{n - \lceil pn/q \rceil} b^{\lceil pn/q \rceil - 1}.$$

Putting $n = sq + r$, $s \ge 0, 0 \le r < q$, we have

$$\frac{1}{|K_1^*|} = -\frac{1}{b} + \sum_{s=0}^{\infty} (a^{q-p}b^p)^s \sum_{r=0}^{q-1} a^{r - \lceil pr/q \rceil} b^{\lceil pr/q \rceil - 1},$$

which is equal to

$$-\frac{1}{b}+\frac{1}{1-a^{q-p}b^{p}}\left(P^{+}_{p,q}\left(a,\frac{b}{a}\right)+\frac{1}{b}-a^{q-p}b^{p-1}\right)=\frac{P^{+}_{p,q}(a,b/a)}{1-a^{q-p}b^{p}},$$

because $a^{q-p}b^{p}<1$ by Lemma 8.1. Since $|K_1^*|=1-b-(a-b)c$, this completes the proof. □

By taking account of only the intervals K_n^* lying in $[c,1)$, we get

Theorem 8.4. *For any $\psi_{a,b,c}\in\mathfrak{E}^0(p/q)$ we have*

$$P_{q,p}\left(\frac{b}{a},a\right)=\frac{a(1-c)(1-a^{q-p}b^{p})}{1-b-(a-b)c}.$$

Proof. Put $\psi=\psi_{a,b,c}$ for brevity. In the same way as the proof of Theorem 8.2 we have

$$\begin{aligned}1-c&=\sum_{\substack{n\ge1\\ \psi^n(1)>c}}\left|K_n^*\right|\\ &=\sum_{n=1}^{\infty}\epsilon\circ\psi^n(1)\left|K_n^*\right|.\end{aligned}$$

Since $\epsilon\circ\psi(1)=\mu_1$ and

$$\begin{aligned}\epsilon\circ\psi^n(1)&=\mu_n-\mu_{n-1}\\ &=\lceil p(n+1)/q\rceil-\lceil pn/q\rceil,\end{aligned}$$

we get

$$\frac{1-c}{|K_1^*|}=\sum_{n=1}^{\infty}\left(\left\lceil\frac{p}{q}(n+1)\right\rceil-\left\lceil\frac{p}{q}n\right\rceil\right)a^{n-\lceil pn/q\rceil}b^{\lceil pn/q\rceil-1}.$$

Adjusting the term $1/b$ corresponding to $n=0$ and putting $n=sq+r$, $s\ge0$, $0\le r<q$, we have

$$\frac{1-c}{|K_1^*|}=-\frac{1}{b}+\sum_{s=0}^{\infty}(a^{q-p}b^{p})^{s}\sum_{r=0}^{q-1}\left(\left\lceil\frac{p}{q}(r+1)\right\rceil-\left\lceil\frac{p}{q}r\right\rceil\right)a^{r-\lceil pr/q\rceil}b^{\lceil pr/q\rceil-1}.$$

Since $0<p/q<1$, $\lceil pr/q\rceil$ takes all integers from 0 to p as $r\in[0,q)$ varies. Putting $\lceil p(r+1)/q\rceil=\ell+1$ and $\lceil pr/q\rceil=\ell$, we have $pr/q\le\ell<p(r+1)/q$, and so $r=[q\ell/p]$. Here it is worthwhile to note that ℓ varies only from 0 up to $p-1$,

because $r < q$. Hence

$$\begin{aligned}\frac{(1-c)(1-a^{q-p}b^{p})}{|K_1^*|} &= -\frac{1}{b} + a^{q-p}b^{p-1} + \sum_{\ell=0}^{p-1} a^{[q\ell/p]-\ell}b^{\ell-1} \\ &= \frac{1}{a}P_{q,p}\left(\frac{b}{a}, a\right).\end{aligned}$$

Since $|K_1^*| = 1 - b - (a-b)c$, this completes the proof. □

8.2 Limit Sets Ω_ψ

Theorem 8.5. *The limit set* Ω_ψ *for* $\psi = \psi_{a,b,c} \in \mathfrak{B}_\infty$ *is a null set*‡ *and totally disconnected.*

Proof. Making the same argument as in Section 6.2 under the assumption that $c_{2-n} \in \operatorname{Int} \psi(J_1)$, it is verified that $\epsilon \circ \psi^k(0) = \epsilon \circ g^k(0)$ for $1 \le k \le n-2$ and

$$\sum_{k=1}^{n-2} \epsilon \circ \psi^k(0) = p - 1, \tag{8.1}$$

because $g^{n-1}(0) = c$ and $g \in \mathfrak{E}^0(p/n)$. Since the restriction $\psi^{n-1}|J_1$ is a linear function with slope $a^{n-p}b^{p-1}$, we have

$$\left|\psi^{n-1}(J_1)\right| = a^{n-p}b^{p-1}\,|J_1|.$$

On the other hand, each interval J_k, $2 \le k \le n$ can be expressed as $[c_{-s}, *)$ for some $0 \le s \le n$. Since $\psi^j(c_{-s}) = g^j(c_{-s})$ for $0 \le j \le s+1$ and since

$$\begin{aligned}\epsilon \circ \psi^j(c_{-s}) &= \epsilon \circ \psi^{j-s-1}(0) \\ &= \epsilon \circ g^{j-s-1}(0) \\ &= \epsilon \circ g^j(c_{-s})\end{aligned}$$

for $s+1 < j \le n-2$, the digits $\epsilon \circ \psi^j(c_{-s})$, $0 \le j \le n-2$ are $n-1$ consecutive one's in the periodic itinerary by g with period n. Therefore

$$p - 1 \le \sum_{j=0}^{n-2} \epsilon \circ \psi^j(c_{-s}) \le p,$$

and hence

$$\left|\psi^{n-1}(J_k)\right| \le \max\left(\frac{1}{a}, \frac{1}{b}\right) a^{n-p}b^{p}\,|J_k|.$$

‡ A measurable set E is called a null set if $|E| = 0$.

Using the inequality

$$\rho(n-1) \leq p \leq \rho(n-1)+1$$

from (8.1) where $\rho = \rho(\psi)$, we thus have

$$\left|\psi^{n-1}(J)\right| \leq \max\left(\frac{1}{a}, \frac{1}{b}, \frac{b}{a}, \frac{a}{b}\right)(a^{1-\rho}b^{\rho})^{n-1}. \tag{8.2}$$

We next assume that $c_{2-n} \in \operatorname{Int}\psi(J_n)$. In this case $\epsilon \circ \psi^k(0) = \epsilon \circ h^k(0)$ for $1 \leq k \leq n-2$ and hence

$$\sum_{k=1}^{n-2} \epsilon \circ \psi^k(0) = p - 1,$$

because $h^{n-1}(1) = c, h \in \mathfrak{E}^{\phi}(p/n)$ and $\eta \circ h(0) = \eta \circ h(1)$ by (i) of Theorem 4.6. Since the restriction $\psi^{n-1} \mid J_n$ is a linear function with slope $a^{n-p}b^{p-1}$, we have

$$\left|\psi^{n-1}(J_n)\right| = a^{n-p}b^{p-1} \, |J_n|.$$

For other interval J_k, $1 \leq k < n$ can be expressed as $[*, c_{-t})$ for some $0 \leq t \leq n$. Since $\psi^j(c_{-t}) = h^j(c_{-t})$ for $0 \leq j \leq t+1$ and since

$$\begin{aligned}
\epsilon \circ \psi^j(c_{-t}) &= \epsilon \circ \psi^{j-t-1}(0) \\
&= \epsilon \circ h^{j-t-1}(0) \\
&= \epsilon \circ h^j(c_{-t})
\end{aligned}$$

for $t+1 < j \leq n-2$, the digits $\epsilon \circ \psi^j(c_{-t})$, $0 \leq j \leq n-2$ are $n-1$ consecutive one's in the periodic itinerary by h with period n. Therefore

$$p - 1 \leq \sum_{j=0}^{n-2} \epsilon \circ \psi^j(c_{-t}) \leq p,$$

and hence

$$\left|\psi^{n-1}(J_k)\right| \leq \max\left(\frac{1}{a}, \frac{1}{b}\right) a^{n-p}b^{p} \, |J_k|.$$

Thus we obtain the same inequality (8.2). By Lemma 6.2 Case (a) occurs for infinitely many n's, and the same is the case with the inequality (8.2). Thus we conclude that

$$\lim_{n\to\infty} |\psi^n(J)| = 0,$$

because of $a^{1-\rho}b^{\rho} < 1$ by Lemma 8.1 and of the monotonicity of the sequence $|\psi^n(J)|$. Therefore we have $|\Omega_\psi| = 0$, as required. □

Corollary 8.1. *For any* $\psi = \psi_{a,b,c} \in \mathfrak{B}$ *we have* $\rho(\psi) \in (0,1)$. *Moreover* $\rho(\psi) \notin \mathbb{Q}$ *if and only if* $\psi \in \mathfrak{B}_\infty$.

8.3 Basic Formulae of the Second Kind

For an arbitrarily irrational $\mu > 0$ we define the double series

$$M_\mu(w,z) = \sum_{\substack{m,n \ge 1 \\ m < \mu n}} w^m z^n,$$

known as the *Hecke-Mahler series*. The transcendence of Mahler functions including the Hecke-Mahler series is developed in detail by Nishioka (1996).

Note that $M_\mu(w,z)$ converges absolutely if $|z| < 1$ and $|w|^\mu |z| < 1$. Moreover

$$M_\mu(1,z) = \sum_{n=1}^{\infty} [\mu n] z^n = H_\mu(z) \tag{8.3}$$

is the power series defined in (6.6). Since it follows that

$$\begin{aligned} M_\mu(w,z) &= \sum_{m=1}^{\infty} w^m \sum_{n > [m/\mu]} z^n \\ &= \frac{z}{1-z} \sum_{m=1}^{\infty} w^m z^{[m/\mu]}, \end{aligned} \tag{8.4}$$

we have

$$M_\mu(w,z) = \frac{wz}{1-z} Q_{1/\mu}(w,z), \tag{8.5}$$

where

$$Q_{1/\mu}(w,z) = \sum_{m=1}^{\infty} w^{m-1} z^{[m/\mu]} \tag{8.6}$$

converges absolutely in a wider region $|w|^\mu |z| < 1$ so that $z = 1$ is a simple pole of $M_\mu(w,z)$ with residue $w/(w-1)$ for each fixed $0 < |w| < 1$. The formula (8.6) allows us to define $Q_{1/\mu}(w,z)$ even when μ is rational. Indeed, if $1/\mu = p/q$, then we have

$$Q_{p/q}(w,z) = \frac{P_{p,q}(w,z)}{1 - w^q z^p}.$$

Moreover, the analytic function $Q_{1/\mu}(w,z)$ can be considered as the limit of the polynomials $P_{p,q}(w,z)$ as $p, q \to \infty$ and $p/q \to 1/\mu$. To be more exact,

Lemma 8.2. *Let μ be a positive irrational number and let $\{a_n/b_n\}$ be a sequence of rational numbers converging to $1/\mu$. Then the sequence of polynomials $P_{a_n,b_n}(w,z)$ converges uniformly to $Q_{1/\mu}(w,z)$ on any compact subset in the region $|w|^\mu|z| < 1$.*

Proof. Let K be a compact subset in the region $|w|^\mu|z| < 1$. Then, for some small $0 < \varepsilon < \{1/\mu\}$,

$$\lambda = \sup_{\substack{(w,z)\in K \\ |x-1/\mu|<\varepsilon}} |w||z|^x < 1.$$

Let n_0 be an integer with

$$\left|\frac{1}{\mu} - \frac{a_n}{b_n}\right| < \frac{\varepsilon}{2}$$

for all $n \geq n_0$. For any integer N satisfying

$$N > 2\left(1 + \frac{1}{\varepsilon}\max\left(1, \frac{1}{\mu}\right)\right),$$

there exists an integer $M \geq n_0$ in such a way that $\{1/\mu\}$ and the fractional part of a_n/b_n belong to the same Farey interval of order N for all $n \geq M$. Since $[1/\mu] = [a_n/b_n]$, it follows from Lemma 5.5 that

$$\begin{aligned}\left[\frac{k}{\mu}\right] &= \left[\frac{1}{\mu}\right]k + \left[\left\{\frac{1}{\mu}\right\}k\right] \\ &= \left[\frac{a_n}{b_n}\right]k + \left[\left\{\frac{a_n}{b_n}\right\}k\right] = \left[\frac{a_n}{b_n}k\right]\end{aligned}$$

for $1 \leq k \leq N$. Moreover it is easily verified that

$$\left|\frac{1}{\mu} - \frac{[m/\mu]}{m-1}\right| < \varepsilon \quad \text{and} \quad \left|\frac{1}{\mu} - \frac{[a_n m/b_n]}{m-1}\right| < \varepsilon$$

for all $m > N$. Hence

$$\begin{aligned}\left|Q_{1/\mu}(w,z) - Q_{a_n/b_n}(w,z)\right| &\leq \sum_{m>N} |w|^{m-1}\left(|z|^{[m/\mu]} + |z|^{[a_n m/b_n]}\right) \\ &\leq 2\sum_{m>N} \lambda^{m-1} \\ &= \frac{2\lambda^N}{1-\lambda}\end{aligned}$$

for any $n \geq M$ and $(w,z) \in K$. Putting $C_0 = 2(1-\lambda)^{-1}$, we have

$$\left|Q_{1/\mu}(w,z) - P_{a_n,b_n}(w,z)\right| \leq \left|w^{b_n}z^{a_n}\right| \cdot \left|Q_{1/\mu}(w,z)\right| + C_0\lambda^N\left(1 + \left|w^{b_n}z^{a_n}\right|\right)$$

and hence

$$\limsup_{n\to\infty} \left| Q_{1/\mu}(w,z) - P_{a_n,b_n}(w,z) \right| \le C_0 \lambda^N,$$

because $|w^{b_n} z^{a_n}| \le \lambda^{b_n} \to 0$ as $n \to \infty$. Since N is arbitrary, this completes the proof. □

It is worthwhile to see that the sequence of polynomials $P^+_{a_n,b_n}(w,z)$ plays a role similar to $P_{a_n,b_n}(w,z)$ because of the formula stated in Exercise 2 in this chapter.

The following theorem is called the basic formula of the second kind. Compare with Theorem 8.1.

Theorem 8.6. *For any $\psi_{a,b,c} \in \mathfrak{B}_\infty$ we have*

$$Q_\rho\left(a, \frac{b}{a}\right) = \frac{1}{1 - b - (a-b)c}$$

where $\rho = \rho(\psi_{a,b,c})$.

Proof. It follows from Theorem 8.5 that

$$\sum_{n=1}^{\infty} |K_n| = 1.$$

Since

$$|K_{n+1}| = a^{n-\nu_n} b^{\nu_n} |K_1|$$

with $\nu_n = [\rho(n+1)]$, we get

$$\frac{1}{|K_1|} = \sum_{n=1}^{\infty} a^{n-1-[\rho n]} b^{[\rho n]} = Q_\rho\left(a, \frac{b}{a}\right).$$

This completes the proof. □

We also have another basic formula of the second kind by taking account of only the intervals K_n lying in $[c, 1)$.

Theorem 8.7. *For any $\psi_{a,b,c} \in \mathfrak{C}_\infty$ we have*

$$Q_{1/\rho}\left(\frac{b}{a}, a\right) = \frac{a(1-c)}{1 - b - (a-b)c}$$

where $\rho = \rho(\psi_{a,b,c})$.

Proof. The proof is similar to that of Theorem 8.2. We have

$$\begin{aligned} 1 - c &= \sum_{\substack{n \geq 1 \\ \psi^n(0) > c}} |K_n| \\ &= \sum_{n=1}^{\infty} \epsilon \circ \psi^n(0) \, |K_n| \\ &= |K_1| \sum_{n=1}^{\infty} ([\rho(n+1)] - [\rho n]) a^{n-1-[\rho n]} b^{[\rho n]}. \end{aligned}$$

Since $0 < \rho < 1$, $[\rho n]$ take all the non-negative integral values as $n \geq 1$ varies. So, putting $[\rho(n+1)] = \ell$ and $[\rho n] = \ell - 1$, we have $\rho n < \ell < \rho(n+1)$. Therefore $n = [\ell/\rho]$ and so,

$$\begin{aligned} \frac{1-c}{|K_1|} &= \sum_{\ell=1}^{\infty} a^{[\ell/\rho]-\ell} b^{\ell-1} \\ &= \frac{1}{a} Q_{1/\rho}\left(\frac{b}{a}, a\right), \end{aligned}$$

as required. The last series converges absolutely by Lemma 8.1. This completes the proof. □

8.4 Hausdorff Dimension of Ω_ψ

In this section we introduce the Hausdorff dimension, which is an important notion in the theory of fractal sets, giving a way to measure the size of a set. For the sake of simplicity, we deal with only subsets of $\mathbb{R}^n$. We denote by $|x - y|$ the usual Euclidean distance between two points x, y in $\mathbb{R}^n$. For basic notions on measures and some interesting topics on fractal sets, see Falconer (1985), for example.

For a non-empty bounded subset E of $\mathbb{R}^n$ put

$$|E| = \sup_{x, y \in E} |x - y|,$$

which we call the *diameter* of the set E. We say that a sequence of subsets $\{E_k\}_{k=1}^{\infty}$ is a δ-*covering* of $X \subset \mathbb{R}^n$ provided that $0 < |E_k| \leq \delta$ and

$$X \subset \bigcup_{k=1}^{\infty} E_k.$$

For any non-negative number s and a subset $X \subset \mathbb{R}^n$ we then define

$$\mathscr{H}_\delta^s(X) = \inf \sum_{k=1}^{\infty} |E_k|^s,$$

where the infimum in the right-hand side extends over all δ-coverings $\{E_k\}$ of X. It is known that $\mathscr{H}_\delta^s$ is an outer measure on $\mathbb{R}^n$.

For a fixed subset X we can regard the outer measure $\mathscr{H}_\delta^s$ as a function of two variables s and δ, although $\mathscr{H}_\delta^s$ may take the value $+\infty$. It is straightforward to see that $\mathscr{H}_\delta^s$, as a function of δ, is monotonically decreasing. Then we define the *Hausdorff s-dimensional outer measure* of X by

$$\mathscr{H}^s(X) = \lim_{\delta \to 0+} \mathscr{H}_\delta^s(X) \in [0, \infty].$$

Furthermore, $\mathscr{H}^s$ is monotonically decreasing as a function of s, and it can be seen that

$$\sup\{s > 0 : \mathscr{H}^s(X) = \infty\} = \inf\{s > 0 : \mathscr{H}^s(X) = 0\},$$

which is called the *Hausdorff dimension* of X and denoted by $\dim_{\mathrm{H}} X$.

To investigate the Hausdorff dimension of the limit set Ω_ψ we need the following theorem due to Besicovitch and Taylor (1954), which gives an upper bound of the Hausdorff dimension of Cantor-type sets in $\mathbb{R}$. We give the proof for the sake of completeness.

Theorem 8.8. *Suppose that $\{E_k\}_{k=1}^{\infty}$ is a sequence of disjoint open subintervals in a bounded closed interval W. Put*

$$\varepsilon_m = \sum_{k=m}^{\infty} |E_k|$$

for all $m \geq 1$ and suppose that $\varepsilon_1 = |W|$. Then the Hausdorff dimension of the null set

$$X = W \setminus \bigcup_{k=1}^{\infty} E_k$$

is less than or equal to

$$\liminf_{m \to \infty} \frac{\log m}{\log(m/\varepsilon_m)}. \tag{8.7}$$

Proof. For an arbitrarily fixed integer $m \geq 2$ the set difference

$$D_m = W \setminus \bigcup_{k=1}^{m-1} E_k$$

consists of m closed intervals, say $W_1, \ldots, W_m$ from left to right. Since

$$|W_k| \le |D_m| = |W| - \sum_{i=1}^{m-1} |E_i| = \varepsilon_m$$

for $1 \le k \le m$, the finite sequence $\{W_1, \ldots, W_m\}$ becomes an ε_m-covering of X if none of them consist of only a single point. However, even so, we can enlarge them slightly so that the new sequence is still a $2\varepsilon_m$-covering of X.

Let λ be the infimum in (8.7). We can assume that $\lambda < 1$. For any $\epsilon \in (0, 1-\lambda)$ there exists a subsequence of integers $\{m_j\}$ satisfying

$$m_j \left(\frac{\varepsilon_{m_j}}{m_j} \right)^{\lambda+\epsilon} < 1.$$

Since the function $x^{\lambda+\epsilon}$ is concave downward, we have

$$\frac{1}{m_j} \sum_{k=1}^{m_j} |W_k^{m_j}|^{\lambda+\epsilon} \le \left(\frac{1}{m_j} \sum_{k=1}^{m_j} |W_k^{m_j}| \right)^{\lambda+\epsilon} = \left(\frac{|D_{m_j}|}{m_j} \right)^{\lambda+\epsilon} = \left(\frac{\varepsilon_{m_j}}{m_j} \right)^{\lambda+\epsilon} < \frac{1}{m_j}.$$

Hence

$$\mathscr{H}^{\lambda+\epsilon}_{2\varepsilon_{m_j}}(X) < 1,$$

from which we get $\mathscr{H}^{\lambda+\epsilon}(X) \le 1$ by letting $j \to \infty$. Therefore $\dim_{\mathrm{H}} X \le \lambda + \epsilon$. Since ϵ is arbitrary, this completes the proof. □

For the limit set Ω_ψ for $\psi = \psi_{a,b,c} \in \mathfrak{B}_\infty$ we take $W = [0, 1]$ and $E_k = \mathrm{Int}\, K_k$ for all $k \ge 1$. We then have

$$\varepsilon_m = \sum_{k=m}^{\infty} |K_k| \le \sum_{k=m}^{\infty} a^{k-1-[\rho k]} b^{[\rho k]} \le \max\left(\frac{1}{ab}, \frac{1}{b} \right) \frac{(a^{1-\rho} b^{\rho})^m}{1 - a^{1-\rho} b^{\rho}}$$

for all $m \ge 1$, which implies that

$$\lim_{m \to \infty} \frac{\log m}{\log(m/\varepsilon_m)} = 0.$$

Thus, we have the following

Theorem 8.9. *For any* $\psi = \psi_{a,b,c} \in \mathfrak{B}_\infty$ *we have* $\dim_{\mathrm{H}} \Omega_\psi = 0$.

In comparison with Cantor's ternary set, this set Ω_ψ is fairly thin, because it is well-known that the Hausdorff dimension of Cantor's ternary set is $\log 2/\log 3$, whose Hausdorff $\log 2/\log 3$-dimensional measure is 1.

Exercises in Chapter 8

1 Let p, q be coprime positive integers. Show that

$$P_{p,q}(w,z) - w^{q-1}z^p = P^+_{p,q}(w,z) - w^{q-1}z^{p-1}.$$

[Note that $w^{q-1}z^p$ and $w^{q-1}z^{p-1}$ are the last terms in the definitions of $P_{p,q}$ and $P^+_{p,q}$ respectively.]

2 Let p, q be coprime positive integers. Show the following identity:

$$(1-w)P_{p,q}(w,z) + (1-z)P^+_{q,p}(z,w) = 1 - w^q z^p.$$

3 Let μ be a positive irrational number. Show that the functional relations

$$M_\mu(w,z) + M_{1/\mu}(z,w) = \frac{wz}{(1-w)(1-z)}$$

and

$$(1-w)Q_{1/\mu}(w,z) + (1-z)Q_\mu(z,w) = 1$$

hold for $|w|^\mu |z| < 1$.

4 Let μ be a positive irrational number. Show that

$$\begin{aligned} H_\mu(z) &= \sum_{n=1}^{\infty} [\mu n] z^n \\ &= \frac{z}{1-z} \sum_{n=1}^{\infty} z^{[n/\mu]} \end{aligned}$$

for $|z| < 1$. Derive from this that $H_\mu(1/m) \notin \mathbb{Q}$ for each integer $m \geq 2$.

5 Let μ be a positive irrational number. Show that

$$Q_\mu(w,z) = z^{[\mu]} Q_{\{\mu\}}(wz^{[\mu]}, z)$$

for $|z|^\mu |w| < 1$.

6 The quadratic irrational $\varphi = (\sqrt{5}+1)/2 = 1.61803...$ is known as the *golden ratio*. Show that

$$Q_\varphi(w,z) = \frac{z}{1-wz} - \frac{z(1-z)}{1-wz} Q_\varphi(z, wz)$$

for $|wz| < 1$.

7 Deduce from 6 that

$$Q_\varphi(w,z) = \frac{z}{1-wz} + (1-z)\sum_{n=1}^{\infty}(-1)^n \frac{w^{F_{n+2}-1}z^{F_{n+3}-1}}{(1-w^{F_n}z^{F_{n+1}})(1-w^{F_{n+1}}z^{F_{n+2}})}$$

for $|wz| < 1$ and $|z| < 1$, where F_n is the nth *Fibonacci number* defined by the linear recurrence relation $F_{n+2} = F_{n+1} + F_n$ with $F_1 = F_2 = 1$.

8 Show that

$$H_{1/\varphi}(z) = \sum_{n=1}^{\infty}(-1)^{n-1}\frac{z^{F_{n+2}}}{(1-z^{F_n})(1-z^{F_{n+1}})} = \sum_{n=1}^{\infty}\frac{z^{F_{2n+1}}}{(1-z^{F_{2n-1}})(1-z^{F_{2n+1}})}$$

for $|z| < 1$, where φ is the golden ratio. Deduce from the last formula that the number of triples of positive integral solutions (i, j, k) of the equation

$$n = (i-1)F_{2k-1} + jF_{2k+1}$$

is exactly $[n/\varphi]$ for all $n \geq 1$.

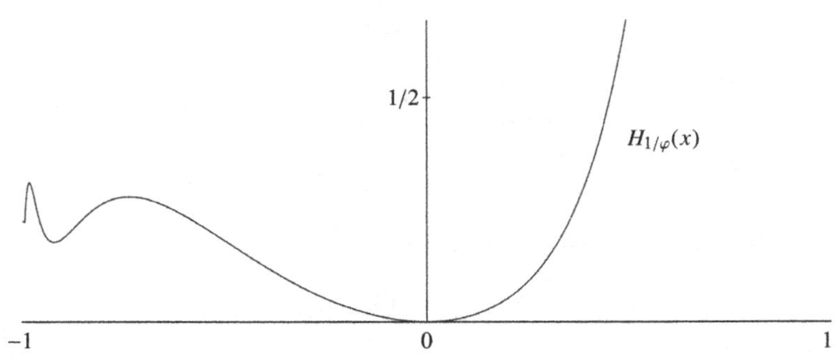

Fig. 8.1 The graph of $H_{1/\varphi}(x)$ on the interval $(-1, 1)$ is plotted. Although the accuracy of graph in the right-neighborhood of $x = -1$ may be unreliable, it will be shown in Section 12.5 that the function indeed oscillates infinitely often.

9 When $a = b$ and $\psi_{a,a,c} \in \mathfrak{B}_\infty$, show that

$$c = 1 - \left(\frac{1-a}{a}\right)^2 H_\rho(a)$$

where $\rho = \rho(\psi_{a,a,c})$.

10 When $a = 1$ and $\psi_{1,b,c} \in \mathfrak{B}_\infty$, show that

$$c = 1 - \frac{b}{(1-b)^2} \cdot \frac{1}{H_{1/\rho}(b)}$$

where $\rho = \rho(\psi_{1,b,c})$.

11 Let α and β be positive irrational numbers satisfying $1/\alpha + 1/\beta = 1$. Show that

$$\sum_{n=1}^{\infty} z^{[\alpha n]} + \sum_{n=1}^{\infty} z^{[\beta n]} = \frac{z}{1-z}$$

for $|z| < 1$.

12 Let μ be a positive irrational number and $P(x)$ be a polynomial satisfying $P(0) = 0$. Show then that

$$\sum_{n=1}^{\infty} P([\mu n]) z^n = \frac{z}{1-z} \sum_{n=1}^{\infty} (P(n) - P(n-1)) z^{[n/\mu]}.$$

Chapter 9

Orbital and Itinerary Functions

The orbits of $\psi_{a,b,c}$ starting from 0 and 1 are regarded as the set of functions of variable c and are called the "orbital" functions. The discontinuity points of the orbital functions are investigated, which play an important role in the next chapter.

9.1 Orbital Functions u_n and v_n

Recall that the set of parameters (a, b, c) satisfying $\psi_{a,b,c} \in \mathfrak{B}(c)$ is

$$\Lambda = \left\{(a, b, c) \in \mathbb{R}^3 : a, b > 0,\ ac + b(1 - c) < 1,\ 0 < c < 1\right\}.$$

However we can assume that $a \geq b$; otherwise consider the reflection of $\psi_{a,b,c}$, which is equal to $\psi_{b,a,1-c}$ (see Exercise 2 in Chapter 3). So the parameter region for the piecewise linear functions treated from now on is

$$\Lambda_0 = \left\{(a, b, c) \in \mathbb{R}^3 : a \geq b,\ 0 < b < 1,\ 0 < c < r_{a,b}\right\},$$

where

$$r_{a,b} = \begin{cases} 1 & (0 < a \leq 1), \\ \dfrac{1 - b}{a - b} & (a > 1). \end{cases}$$

For given a, b satisfying $a \geq b, 0 < b < 1$ we define

$$u_n(c) = \psi^n_{a,b,c}(0) \quad \text{and} \quad v_n(c) = \psi^n_{a,b,c}(1)$$

as functions of $c \in (0, r_{a,b})$ for $n \geq 1$. We call u_n and v_n the *orbital functions*. Note that the graphs of u_n and v_n do not intersect on $(0, r_{a,b})$ because of the injectivity of $\psi^n_{a,b,c}$. For example, we have

$$u_1(c) = 1 - ac > v_1(c) = b(1 - c).$$

First of all, we need some basic properties of the orbital functions.

Lemma 9.1. *For any $n \geq 1$ the orbital functions u_n and v_n are left-continuous piecewise linear functions with negative slope.*

Proof. We prove by induction on n. Clearly it is true for $u_1(c)$ and $v_1(c)$. We next suppose that the statement holds for some $n \geq 1$. Since $u_n(c)$ is a piecewise linear function with negative slope, the equation $u_n(c) = c$ has only a finite number of solutions. Thus $u_{n+1}(c)$ is also a piecewise linear function with negative slope, because

$$u_{n+1}(c) = \begin{cases} 1 + a(u_n(c) - c) & (u_n(c) < c), \\ b(u_n(c) - c) & (u_n(c) \geq c). \end{cases} \tag{9.1}$$

Moreover, since $\mathrm{H}(u_n(c) - c)$ is left-continuous by assumption, it is easily seen that $u_{n+1}(c)$ is also left-continuous. Also $v_n(c)$ possesses the same properties, because it satisfies the same recurrence relation as (9.1). □

It may be convenient to define

$$u_n(r_{a,b}) = u_n(r_{a,b}-) \quad \text{and} \quad v_n(r_{a,b}) = v_n(r_{a,b}-)$$

so that u_n and v_n are left-continuous piecewise linear functions with negative slope defined on the interval $(0, r_{a,b}]$.

Lemma 9.2. *For any $n \geq 1$ we have*

(i) $u_n(0+) = b^{n-1}$, $v_n(0+) = b^n$,

(ii) *If $0 < a \leq 1$, then $u_n(1) = 1 - a^n$, $v_n(1) = 1 - a^{n-1}$,*

(iii) *If $a > 1$, then*

$$u_n(r_{a,b}) = v_n(r_{a,b}) = \frac{a^i b^j - b}{a - b},$$

where $i = n - [\kappa n]$ and $j = 1 + [\kappa n]$ with

$$\kappa = \kappa_{a,b} = \frac{\log a}{\log(a/b)} \in (0, 1).$$

Proof. (i) and (ii) will be easily verified by induction on n. We also show (iii) by induction on n. It is true for $n = 1$ because $i = j = 1$. Clearly $ab \in (b, a)$.

Assume next that (iii) holds for some $n \geq 1$ with $i = n - [\kappa n]$, $j = 1 + [\kappa n]$ and $a^i b^j \in (b, a)$. If $a^i b^j \in (b, 1)$, then

$$\begin{aligned} u_n(r_{a,b}-) &= \frac{a^i b^j - b}{a - b} \\ &< \frac{1 - b}{a - b} = r_{a,b}. \end{aligned}$$

Therefore

$$u_{n+1}(r_{a,b}-) = 1 + a(u_n(r_{a,b}-) - r_{a,b})$$
$$= \frac{a^{i+1}b^j - b}{a - b}$$

and $a^{i+1}b^j$ belongs to the interval (b, a). On the other hand, if $a^i b^j \in [1, a)$, then

$$u_n(r_{a,b}-) = \frac{a^i b^j - b}{a - b}$$
$$\geq \frac{1 - b}{a - b} = r_{a,b}.$$

Since the graph of u_n is above the diagonal in a left-neighborhood of $r_{a,b}$, we have

$$u_{n+1}(r_{a,b}-) = b(u_n(r_{a,b}-) - r_{a,b})$$
$$= \frac{a^i b^{j+1} - b}{a - b}$$

and $a^i b^{j+1}$ belongs to the interval $[b, a)$.

We thus have $a^{n+2-m}b^m \in [b, a)$ for some integer m in each case, which implies that $m - 1 \leq \kappa(n+1) < m$. Hence $m = 1 + [\kappa(n+1)]$, as required. □

Note that the sequences $\{u_n(r_{a,b})\}$ and $\{v_n(r_{a,b})\}$ are strictly monotonically increasing if and only if $0 < a < 1$.

9.2 u- and v-Discontinuity Points

Let A_n and B_n be the sets of discontinuity points of $u_n(c)$ and $v_n(c)$ in the interval $(0, r_{a,b})$ respectively. For example, $A_1 = B_1 = \emptyset$.

The following is the first basic lemma on these sets. Although we do not treat the orbital functions as dynamical systems, we may use the notation $\mathrm{Fix}(u_n)$ for the set of all points $c \in (0, r_{a,b})$ satisfying $u_n(c) = c$, ditto with v_n.

Lemma 9.3. *For all $n \geq 1$, we have*

$$A_{n+1} = A_n \cup \mathrm{Fix}(u_n) \quad \textit{and} \quad B_{n+1} = B_n \cup \mathrm{Fix}(v_n).$$

Proof. We give the proof only for u_n, because that for v_n is similar. If $c \notin A_n$ and $u_n(c) \neq c$, then it is obvious from (9.1) that $c \notin A_{n+1}$. Thus the inclusion relation $A_{n+1} \subset A_n \cup \mathrm{Fix}(u_n)$ holds.

Conversely, suppose first that $A_n \not\subset A_{n+1}$. Then there exists $c \in A_n \backslash A_{n+1}$, at which $u_n(c+) \neq u_n(c)$ holds. Since $\psi_{a,b,c}$ is injective, we have

$$\psi_{a,b,c} \circ u_n(c+) \neq \psi_{a,b,c} \circ u_n(c) = u_{n+1}(c).$$

If $u_n(c+) \neq c$, then $u_{n+1}(c+) = \psi_{a,b,c} \circ u_n(c+)$ by (9.1), contrary to $c \notin A_{n+1}$. Therefore we have $u_n(c+) = c$ and the graph of u_n is below the diagonal in a right-neighborhood of c. Since $c \notin A_{n+1}$, it follows that

$$\begin{aligned} u_{n+1}(c) &= u_{n+1}(c+) \\ &= 1 + a(u_n(c+) - c) = 1, \end{aligned}$$

contrary to $u_{n+1}(c) < 1$. Thus the inclusion relation $A_n \subset A_{n+1}$ holds.

Consider next any point $c \in \mathrm{Fix}(u_n) \backslash A_n$. Then the graph of u_n is below the diagonal in a right-neighborhood of c. Hence

$$u_{n+1}(c+) = 1 + a(u_n(c) - c) = 1,$$

while

$$u_{n+1}(c) = b(u_n(c) - c) = 0.$$

Therefore $c \in A_{n+1}$ and we have $\mathrm{Fix}(u_n) \backslash A_n \subset A_{n+1}$, as required. □

The above lemma shows that any discontinuity point of u_n must be that of u_{n+1}, and that A_n coincides with all the division points of u_n as a piecewise linear function. The same goes for v_n.

Define the sets of all u- and v-discontinuity points by

$$A = \bigcup_{n=1}^{\infty} A_n \quad \text{and} \quad B = \bigcup_{n=1}^{\infty} B_n$$

respectively. For any point $c \in A$ there exists a unique integer $k \geq 1$ satisfying $c \notin A_k$ and $c \in \mathrm{Fix}(u_k)$ because of $A_1 = \emptyset$, which we call the u-discontinuity point of order $k + 1$. Similarly we define the v-discontinuity point of order $k + 1$. Any u-discontinuity point of order $k + 1$ is a discontinuity point of u_{k+1}, but a continuity point of u_k. Thus $A_n^\circ = A_n \backslash A_{n-1}$ is the set of all u-discontinuity points of order $n \geq 2$. We denote similarly the set of all v-discontinuity points of order n by $B_n^\circ = B_n \backslash B_{n-1}$. Then

$$A = \bigcup_{n=2}^{\infty} A_n^\circ \quad \text{and} \quad B = \bigcup_{n=2}^{\infty} B_n^\circ$$

are disjoint unions.

Lemma 9.4. *We have* $A \cap B = \emptyset$.

Proof. Suppose, on the contrary, that there exists a point $c \in A_r^\circ \cap B_s^\circ$ for some integers $r, s \geq 2$. This means that $\psi = \psi_{a,b,c}$ satisfies

$$\psi^{r-1}(0) = \psi^{s-1}(1) = c.$$

We thus have $c \in \mathrm{Per}(\psi)$ and so $1 = c_{1-s}$, contrary to the injectivity of ψ on the interval $[0, 1]$. □

9.3 Itinerary Functions U_n and V_n

We define

$$U_n(c) = \mathrm{H}(u_n(c) - c) \quad \text{and} \quad V_n(c) = \mathrm{H}(v_n(c) - c)$$

for all $n \geq 1$, which we call the itinerary functions. Since the orbital functions u_n, v_n are piecewise linear left-continuous functions with negative slope and since the Heaviside step function $\mathrm{H}(x)$ is right-continuous, the itinerary functions U_n, V_n are left-continuous on the interval $(0, r_{a,b}]$. For example,

$$U_1(c) = \begin{cases} 1 & (0 < c \leq (1+a)^{-1}), \\ 0 & ((1+a)^{-1} < c \leq r_{a,b}), \end{cases}$$

if $ab < 1$, and $U_1(c) \equiv 1$ otherwise†.

The following theorem is a straightforward consequence from Theorem 4.3, Lemma 6.3 and Corollary 8.1.

Theorem 9.1. *For any $0 < c < r_{a,b}$ and $n \geq 1$ we have*

$$U_n(c) = [\rho(n+1)] - [\rho n],$$

where $\rho = \rho(\psi_{a,b,c})$. In other words, $\{U_n(c)\}_{n\geq 1} = \mathfrak{S}(\rho)$.

With respect to $\{V_n(c)\}$ we have the following from Lemma 4.3 and Theorem 4.6:

Theorem 9.2. *For any $0 < c < r_{a,b}$ we have*

(i) $\{V_n(c)\}_{n\geq 1} = \mathfrak{S}(\rho)$ *for* $\psi_{a,b,c} \in \mathfrak{B}_\infty \cup \mathfrak{E}^\phi(p/q)$, *where* $\rho = \rho(\psi_{a,b,c})$,

(ii) $\{V_n(c)\}_{n\geq 1} = \mathfrak{S}^*(p/q)$ *for* $\psi_{a,b,c} \in \mathfrak{E}^*(p/q) \cup \mathfrak{E}^0(p/q)$.

Now we are interested in the behavior of the itinerary functions at the end-points of $(0, r_{a,b}]$.

Lemma 9.5. *For all $n \geq 1$ we have*

(a) $U_n(0+) = V_n(0+) = 1$,

---

† The notation "≡" reads as "equals identically on the domain of definition".

(b) $U_n(1) = V_n(1) = 0$ *if* $0 < a \leq 1$, *and*

$$\{U_n(r_{a,b})\}_{n\geq 1} = \{V_n(r_{a,b})\}_{n\geq 1} = \mathfrak{S}(\kappa_{a,b})$$

if $a > 1$.

Proof. Since (a) and the first part of (b) follow from Lemma 9.2 immediately, we assume that $a > 1$. From (iii) of Lemma 9.2 we have

$$\begin{aligned} U_n(r_{a,b}) &= V_n(r_{a,b}) \\ &= \mathrm{H}\left(\frac{a^i b^j - b}{a - b} - \frac{1 - b}{a - b}\right) \\ &= \mathrm{H}(a^i b^j - 1), \end{aligned}$$

which is equal to 0 if and only if $a^i b^j < 1$. We simply write $\kappa_{a,b}$ as κ. Since

$$a^i b^j = a^n b\left(\frac{b}{a}\right)^{\kappa n - \{\kappa n\}} = \left(\frac{a}{b}\right)^{\kappa + \{\kappa n\} - 1},$$

it follows that $a^i b^j < 1$ if and only if $\kappa + \{\kappa n\} < 1$. Therefore we have

$$\begin{aligned} U_n(r_{a,b}) &= V_n(r_{a,b}) \\ &= [\kappa + \{\kappa n\}] \\ &= [\kappa(n+1)] - [\kappa n], \end{aligned}$$

as required. □

Concerning the behavior of the itinerary functions at their discontinuity points, we have

Lemma 9.6. *Let* β *be any point in* B_q° *for some integer* $q \geq 2$. *Then*

$$u_n(\beta) > v_n(\beta) \quad \textit{and} \quad U_n(\beta) = V_n(\beta)$$

hold for all $n \geq 1$. *Moreover the sequence* $\{U_n(\beta)\}_{n\geq 1}$ *is periodic with period* q.

Proof. We write $\psi_{a,b,\beta}$ as ψ. Since $\psi^{q-1}(1) = \beta$, it follows that $\psi^q(1) = 0$, and so, $\psi \in \mathfrak{E}^\phi(p/q)$ for some irreducible fraction p/q by (i) of Theorem 4.7. From Theorem 4.5 we have

$$u_n(\beta) = \psi^n(0) > \psi^n(1) = v_n(\beta)$$

for all $n \geq 1$. Moreover $U_n(\beta) = V_n(\beta)$ for all $n \geq 1$ by (i) of Theorem 9.2. The sequence $\{U_n(\beta)\}$ is periodic with period q from Theorem 4.2. □

Lemma 9.7. *Let α be any point in A_s° for some integer $s \geq 2$. Then*

$$u_n(\alpha+) > v_n(\alpha+) \quad \text{and} \quad U_n(\alpha+) = V_n(\alpha+) = V_n(\alpha)$$

hold for all $n \geq 1$. Moreover the sequence $\{U_n(\alpha+)\}_{n\geq 1}$ is periodic with period s and is the tail inversion sequence of $\{U_n(\alpha)\}_{n\geq 1}$. In particular, $U_{s-1}(\alpha+) = 0$ and $U_s(\alpha+) = 1$.

Proof. We write $\psi_{a,b,\alpha}$ as ψ. Since α is a continuity point of each $u_1, \ldots, u_{s-1}$, it follows that $\alpha \notin \mathrm{Fix}(u_i)$ for $1 \leq i < s-1$ and $\alpha \in \mathrm{Fix}(u_{s-1})$. This means that $\psi^s(0) = 0$ and $0 \in \mathrm{Per}_s(\psi)$, and hence $\psi \in \mathfrak{C}^0(r/s)$ for some irreducible fraction r/s by (ii) of Theorem 4.7. Since $\alpha \notin B$ by Lemma 9.4, α is a continuity point of v_k and so $\alpha \notin \mathrm{Fix}(v_k)$ for all $k \geq 1$. Thus $u_i(\alpha+) = u_i(\alpha)$ for $1 \leq i \leq s-1$ and $v_k(\alpha+) = v_k(\alpha)$ for all $k \geq 1$.

From (i) of Lemma 4.2 we have

$$u_i(\alpha) = \psi^i(0) > \psi^i(1) = v_i(\alpha)$$

for $1 \leq i < s$, which implies that $u_n(\alpha+) > v_n(\alpha+)$ for $1 \leq n \leq s-1$. Moreover we have $u_s(\alpha+) = 1 > v_s(\alpha)$ and

$$\begin{aligned} u_{s+k}(\alpha+) &= v_k(\alpha) \\ &= \psi^k(1) \\ &> \psi^{s+k}(1) = v_{s+k}(\alpha) \end{aligned}$$

for all $k \geq 1$ by (ii) of Theorem 4.4, which implies that $u_n(\alpha+) > v_n(\alpha+)$ for all $n \geq s$.

On the other hand, $V_n(\alpha) = V_n(\alpha+)$ forms a periodic sequence with period s and it follows from (ii) of Theorem 9.2 that

$$U_i(\alpha+) = U_i(\alpha) = V_i(\alpha)$$

for $1 \leq i \leq s-2$. Since $U_{s-1}(\alpha) = 1$ and $U_s(\alpha) = 0$ by Theorem 9.1, we have $V_{s-1}(\alpha) = 0$ and $V_s(\alpha) = 1$ by (ii) of Theorem 9.2. Moreover the function $u_{s-1}(c)$ comes across the diagonal with negative slope at $c = \alpha$. Thus $U_{s-1}(\alpha+) = 0$ and $U_s(\alpha+) = 1$. Since

$$U_{s+k}(\alpha+) = V_k(\alpha) = V_{s+k}(\alpha)$$

for all $k \geq 1$, it follows that $U_n(\alpha+) = V_n(\alpha+)$ for all $n \geq 1$. This completes the proof. □

Lemma 9.8. *Let $0 \leq \alpha < \beta \leq r_{a,b}$. If $U_k(\alpha+) = U_k(\beta)$ for $1 \leq k \leq n$, then the function $u_{n+1}(c)$ is continuous in the interval $(\alpha, \beta]$. Similarly, if $V_k(\alpha+) = V_k(\beta)$ for $1 \leq k \leq m$, then the function $v_{m+1}(c)$ is continuous in the interval $(\alpha, \beta]$.*

Proof. Let γ be a u-discontinuity point in the interval (α, β) having the smallest order, say ℓ. Then $u_{\ell-1}$ is continuous on (α, β) and $u_{\ell-1}(\gamma) = \gamma$ holds. Therefore

$$U_{\ell-1}(\alpha+) = 1 \quad \text{and} \quad U_{\ell-1}(\beta) = 0,$$

which implies that $\ell - 1 > n$. In particular, the function u_{n+1} is continuous on (α, β). The same proof goes for v_{m+1}. □

Exercises in Chapter 9

1 Let $0 < b < 1 < a$. Show that the sequence $\{u_n(r_{a,b})\}$ is periodic if and only if $\kappa_{a,b} \in \mathbb{Q}$.

2 Let $0 < b < 1 < a$. Show that $\rho(\psi_{a,b,c}) > \kappa_{a,b}$ for any $0 < c < r_{a,b}$.

3 For given b, c satisfying $b > 0$, $0 < c < 1$ and $b(1-c) < 1$, show that the rotation number of $\psi_{a,b,c}$ is a decreasing function of a on the interval

$$\left(0, b + \frac{1-b}{c}\right).$$

4 For given a, c satisfying $a > 0$, $0 < c < 1$ and $ac < 1$, show that the rotation number of $\psi_{a,b,c}$ is an increasing function of b on the interval

$$\left(0, \frac{1-ac}{1-c}\right).$$

5 Let $0 < b < 1 < a$ and put

$$\phi(x) = \left\{\frac{a+b}{2}\left(x - \frac{1-b}{a-b}\right) - \frac{a-b}{2}\left|x - \frac{1-b}{a-b}\right|\right\}.$$

Show that the rotation number of ϕ is

$$\kappa_{a,b} = \frac{\log a}{\log(a/b)}.$$

6 Let $a, c \in (0, 1)$.

(i) Show that the orbital average

$$\rho^* = \lim_{n\to\infty} \frac{x + \psi(x) + \cdots + \psi^{n-1}(x)}{n}$$

exists and is independent of the choice of $x \in [0, 1)$, where $\psi = \psi_{a,a,c}$.

(ii) Show that $\rho(\psi_{a,a,c}) + (1-a)\rho^* = 1 - ac$.

(iii) Show that $f(a,c) \leq \rho(\psi_{a,a,c}) \leq g(a,c)$, where

$$f(a,c) = \begin{cases} (1 + a^2 - a(1+a)c)/2 & (0 < c \leq a/(1+a)), \\ (1+a)(1-c)/2 & (a/(1+a) < c < 1), \end{cases}$$

and

$$g(a,c) = \begin{cases} 1 - (1+a)c/2 & (0 < c \leq 1/(1+a)), \\ (1+a)(1-ac)/2 & (1/(1+a) < c < 1). \end{cases}$$

Chapter 10

Farey Structure

Using the Farey fractions we construct a sequence of closed intervals $\Delta(p/q)$ on c-axis for every irreducible fraction p/q in $(0, r_{a,b})$, on which the corresponding map $\psi_{a,b,c}$ belongs to $\mathfrak{E}^{\phi}(p/q)$, $\mathfrak{E}^{*}(p/q)$ or $\mathfrak{E}^{0}(p/q)$. To calculate the distance between two adjoining Δ's we evaluate the following determinants:

$$\begin{vmatrix} P_{p,q}(w,z) & P^{+}_{q,p}(z,w) \\ P^{+}_{p,q}(w,z) & P_{q,p}(z,w) \end{vmatrix} \quad \text{and} \quad \begin{vmatrix} P_{p,q}(w,z) & P^{+}_{q,p}(z,w) \\ P^{+}_{r,s}(w,z) & P_{s,r}(z,w) \end{vmatrix}.$$

(See Lemmas 10.1 and 10.2 respectively.)

The values of c, for which $\psi_{a,b,c}$ belongs to $\mathfrak{B}_{\infty}$, form the residual set $\Gamma_{a,b}$. This set is extremely thin, because its Hausdorff dimension is zero regardless of the parameters.

10.1 Construction when $r_{a,b} = 1$

We first consider the case $a \leq 1$ so that $r_{a,b} = 1$. We deal with intervals in c-axis such that the itinerary functions behave periodically at their endpoints.

We now introduce the proposition

$$\mathfrak{P}\left(\frac{p}{q}, \frac{r}{s}; \alpha, \beta\right),$$

concerning fractions† $p/q, r/s$ in the interval $[0, 1]$ with $qr - ps = 1$, defined by the following two conditions:

(1) There exist $\alpha \in \{0\} \cup A_s^{\circ}$ and $\beta \in \{1\} \cup B_q^{\circ}$ satisfying $\alpha < \beta$ and

$$p = \sum_{k=1}^{q} U_k(\beta).$$

(2) $U_k(\alpha+) = U_k(\beta)$ holds for $1 \leq k < q + s - 1$.

† We interpret $p/q = 0$ and $r/s = 1$ as meaning that $p = 0$, $q = 1$ and $r = s = 1$ respectively.

Note that the sequence $\{U_n(\beta)\}$ in the condition (1) is periodic with period q by Lemma 9.6. To be more precise, it follows from Lemma 9.6 and Theorem 9.1 that

$$U_n(\beta) = \left[\frac{p}{q}(n+1)\right] - \left[\frac{p}{q}n\right]$$

for all $n \geq 1$ if $\beta \in B_q^\circ$. Moreover this formula is correct even if $\beta = 1$ by (ii) of Lemma 9.5. In particular, the proposition

$$\mathfrak{P}\left(\frac{0}{1}, \frac{1}{1}; 0, 1\right)$$

is true, because the condition (2) is vacuously true when $q = s = 1$.

Suppose now that the proposition $\mathfrak{P}(p/q, r/s; \alpha, \beta)$ is true for some $\alpha \in \{0\} \cup A_s^\circ$, $\beta \in \{1\} \cup B_q^\circ$ and for some fractions $p/q, r/s$ in the interval $[0, 1]$ satisfying $qr - ps = 1$.

We continue our consideration step by step as follows:

1st Step: $u_{q+s-1}(c)$ and $v_{q+s-1}(c)$ are continuous on the interval $(\alpha, \beta]$.

The continuity of u_{q+s-1} on $(\alpha, \beta]$ is straightforward by the condition (2) and Lemma 9.8. It also follows from Lemmas 9.6 and 9.7 that $V_k(\alpha+) = V_k(\beta)$ holds for $1 \leq k < q + s - 1$, which implies the continuity of v_{q+s-1} on $(\alpha, \beta]$ by Lemma 9.8.

2nd Step: $\alpha < v_{q+s-1}(\alpha+) < u_{q+s-1}(\alpha+)$.

Since this is clear by (i) of Lemma 9.2 when $\alpha = 0$, we can assume that $\alpha \in A_s^\circ$. If $\beta \in B_q^\circ$, then $q \geq 2$ and

$$\begin{aligned} U_{q+s-1}(\alpha+) &= U_{q-1}(\alpha+) \\ &= U_{q-1}(\beta) \\ &= \left[\frac{p}{q}q\right] - \left[\frac{p}{q}(q-1)\right] = 1 \end{aligned}$$

by Lemma 9.7, the condition (2) and Theorem 9.1. On the other hand, if $\beta = 1$, then $q = 1$ and

$$U_{q+s-1}(\alpha+) = U_s(\alpha+) = 1$$

by Lemma 9.7. In both cases we have $V_{q+s-1}(\alpha+) = 1$ and

$$\alpha \leq v_{q+s-1}(\alpha+) < u_{q+s-1}(\alpha+).$$

Suppose then that $\alpha = v_{q+s-1}(\alpha+)$ holds. Since $\alpha > 0$, we have $v_{q+s-1}(\alpha) = \alpha$ and $\alpha \in \mathrm{Fix}(v_{q+s-1}) \subset B$, contrary to $\alpha \notin B$. Thus we have $\alpha < v_{q+s-1}(\alpha+)$.

3rd Step: $v_{q+s-1}(\beta) < u_{q+s-1}(\beta) < \beta$.

Since this is clear by (ii) of Lemma 9.2 when $\beta = 1$, we then assume that $\beta \in B_q^\circ$. By Lemma 9.6 we have $v_{q+s-1}(\beta) < u_{q+s-1}(\beta)$. Moreover

$$\begin{aligned} U_{q+s-1}(\beta) &= \left[\frac{p}{q}(q+s)\right] - \left[\frac{p}{q}(q+s-1)\right] \\ &= \left[p + r - \frac{1}{q}\right] - \left[p + r - \frac{p+1}{q}\right] = 0, \end{aligned}$$

because $p/q < 1$ and $qr - ps = 1$. This means that $u_{q+s-1}(\beta) < \beta$.

4th Step: From the consideration in the first step we see that both $u_{q+s-1}(c)$ and $v_{q+s-1}(c)$ are linear functions with negative slope on the interval $(\alpha, \beta]$. Therefore their graphs intersect with the diagonal by the considerations in the second and third steps. Let $\gamma > \delta$ be unique points in the interval (α, β) satisfying $u_{q+s-1}(\gamma) = \gamma$ and $v_{q+s-1}(\delta) = \delta$. For simplicity we denote the closed interval $[\delta, \gamma]$ by Δ.

Since $v_{q+s-1}(\delta) = \delta$, it follows that $\psi^{q+s}(1) = 0$ where $\psi = \psi_{a,b,\delta}$. Hence we have $\psi \in \mathfrak{C}^\phi(k/(q+s))$ for some irreducible fraction $k/(q+s)$ by (i) of Theorem 4.7. This means that $\delta \in B_{q+s}^\circ$.

Similarly we have $\psi^{q+s}(0) = 0$ where $\psi = \psi_{a,b,\gamma}$ because of $u_{q+s-1}(\gamma) = \gamma$. The condition (2) and Lemma 9.8 imply also that each function $u_k(c)$ is continuous and its graph does not get across the diagonal on the interval (α, β) for $1 \le k < q + s - 1$. This means that $0 \in \mathrm{Per}_{q+s}(\psi)$ and therefore $\gamma \in A_{q+s}^\circ$.

5th Step: For any $c \in \Delta$ we have $\psi_{a,b,c} \in \mathfrak{B}_{q+s}$.

On the interval Δ the inequality $v_{q+s-1}(c) \le c \le u_{q+s-1}(c)$ holds. This implies that $\psi_{a,b,c} \notin \mathfrak{B}_\infty$ because of $c \in \overline{K}_{q+s-1}$. Hence there exists some integer $n \ge 2$ satisfying $\psi_{a,b,c} \in \mathfrak{B}_n$. It follows from (i) and (ii) of Lemma 4.2 that $c \notin \overline{K_i}$ for $1 \le i \le n-2$, which implies that $n \le q + s$.

Suppose now that $n < q + s$. Since $c \in \overline{K}_{n-1}$, we have $v_{n-1}(c) \le c \le u_{n-1}(c)$. Moreover, since $n - 1 < q + s - 1$, it follows from the condition (2), Lemmas 9.6 and 9.7 that

$$\begin{aligned} V_{n-1}(\alpha+) &= U_{n-1}(\alpha+) \\ &= U_{n-1}(\beta) = V_{n-1}(\beta). \end{aligned}$$

Therefore the graphs of $u_{n-1}(c)$ and $v_{n-1}(c)$ do not meet the diagonal on the interval (α, β), which means that $U_{n-1}(c) = V_{n-1}(c)$ for any $\alpha < c < \beta$. Since $U_{n-1}(c) = 1$ for $c \in \Delta$, we have $v_{n-1}(c) = c$, a contradiction. Therefore $n = q + s$ independently of $c \in \Delta$.

6th Step: $\rho(\psi_{a,b,c}) = \dfrac{p+r}{q+s}$ for any $c \in \varDelta$.

We have $U_{q+s}(c) = 0$ for any $c \in \varDelta$ and put

$$L(c) = \sum_{k=1}^{q+s} U_k(c) = \sum_{k=1}^{q+s-1} U_k(c).$$

We have already noticed that each function $u_k(c)$ is continuous and its graph does not get across the diagonal on the interval (α, β) for $1 \le k < q+s-1$. This implies that $U_k(c)$ is constant on the interval $(\alpha, \beta]$, because $U_k(c)$ is left-continuous. On the other hand, we have $U_{q+s-1}(c) = 1$ on the interval $(\alpha, \gamma]$. Therefore, for $c \in \varDelta$,

$$\begin{aligned} L(c) &= \sum_{k=1}^{q+s-2} U_k(\beta) + U_{q+s-1}(c) \\ &= \sum_{k=1}^{q+s-2} \left(\left[\frac{p}{q}(k+1) \right] - \left[\frac{p}{q} k \right] \right) + 1 \\ &= \left[\frac{p}{q}(q+s-1) \right] + 1 \\ &= p + r, \end{aligned}$$

because $p/q < 1$ and $qr - ps = 1$.

Note that, according as $c = \delta$, $\delta < c < \gamma$ or $c = \gamma$, the piecewise linear map $\psi_{a,b,c}$ belongs to

$$\mathfrak{E}^{\phi}\left(\frac{p+r}{q+s}\right), \quad \mathfrak{E}^{*}\left(\frac{p+r}{q+s}\right) \quad \text{or} \quad \mathfrak{E}^{0}\left(\frac{p+r}{q+s}\right).$$

7th Step: $U_k(\alpha+) = U_k(\delta)$ for $1 \le k < q + 2s - 1$.

It has been shown that $U_k(\alpha+) = U_k(\delta)$ for $1 \le k \le q + s - 1$ in the previous step. If $\alpha = 0$, then we have nothing to do any more, because $s = 1$. Suppose then that $\alpha \in A_s^\circ$. For $k = q + s$, we have

$$U_{q+s}(\alpha+) = U_q(\alpha+) = U_q(\beta) = 0$$

because $s \ge 2$ and $q < q + s - 1$, while $U_{q+s}(\delta) = 0$ obviously.

For $q + s < k < q + 2s - 1$ we have

$$\begin{aligned} U_k(\alpha+) &= U_{k-s}(\alpha+) \\ &= U_{k-s}(\beta) \\ &= U_{k-q-s}(\beta) = U_{k-q-s}(\alpha+). \end{aligned}$$

On the other hand, we have

$$U_k(\delta) = U_{k-q-s}(\delta) = U_{k-q-s}(\alpha+),$$

because $k - q - s < s - 1 < q + s - 1$, and hence $U_k(\alpha+) = U_k(\delta)$, as required.

Last Step: $U_k(\gamma+) = U_k(\beta)$ for $1 \le k < 2q + s - 1$.

It has been shown that $U_k(\gamma+) = U_k(\beta)$ for $1 \le k < q + s - 1$ in the previous step. This holds also for $k = q + s - 1$, because $U_{q+s-1}(\gamma+) = 0$ by $u_{q+s-1}(\gamma) = \gamma$ and because $U_{q+s-1}(\beta) = 0$ by the consideration in the third step. If $\beta = 1$, then we have nothing to do any more, because $q = 1$. So we assume that $\beta \in B_q^\circ$. For $k = q + s$, we have

$$U_{q+s}(\beta) = U_s(\beta) = U_s(\alpha+) = 1$$

because $q \ge 2$ and $s < q + s - 1$, while $U_{q+s}(\gamma+) = 1$ by Lemma 9.7.

For $q + s < k < 2q + s - 1$ we have

$$\begin{aligned} U_k(\beta) &= U_{k-q}(\beta) \\ &= U_{k-q}(\alpha+) \\ &= U_{k-q-s}(\alpha+) = U_{k-q-s}(\beta). \end{aligned}$$

On the other hand, we have

$$U_k(\gamma+) = U_{k-q-s}(\gamma+) = U_{k-q-s}(\beta),$$

because $k - q - s < q - 1 < q + s - 1$, and hence $U_k(\gamma+) = U_k(\beta)$, as required.

The above considerations imply that both the propositions

$$\mathfrak{P}\left(\frac{p}{q}, \frac{p+r}{q+s}; \gamma, \beta\right) \quad \text{and} \quad \mathfrak{P}\left(\frac{p+r}{q+s}, \frac{r}{s}; \alpha, \delta\right)$$

are true. Therefore it follows from Lemma 5.4 that $\mathfrak{P}(p/q, r/s; \alpha, \beta)$ is true for all fractions $p/q, r/s$ in the interval $[0, 1]$ satisfying $qr - ps = 1$.

In each proposition $\mathfrak{P}(p/q, r/s; \alpha, \beta)$ we have constructed the closed interval $\Delta = [\delta, \gamma]$, which we denote by

$$\Delta\left(\frac{p+r}{q+s}\right).$$

We thus have the following

Theorem 10.1. *Suppose that* $0 < b < 1$ *and* $b \le a \le 1$. *For each irreducible fraction* p/q *in the interval* $(0, 1)$ *there exists a closed interval* $\Delta(p/q) \subset (0, 1)$ *satisfying the following properties:*

(i) *If $p/q < r/s$, then $\Delta(p/q) > \Delta(r/s)$.*

(ii) *According as c is the left endpoint, an interior point or the right endpoint of $\Delta(p/q)$, the piecewise linear function $\psi_{a,b,c}$ belongs to*

$$\mathfrak{E}^{\phi}\left(\frac{p}{q}\right), \quad \mathfrak{E}^{*}\left(\frac{p}{q}\right) \quad or \quad \mathfrak{E}^{0}\left(\frac{p}{q}\right).$$

10.2 Construction when $r_{a,b} < 1$

We next consider the case $a > 1$ so that $r_{a,b} < 1$. The behavior of the sequence $\{U_n(c)\}$ at the right endpoint is not as simple as in Section 10.1, because

$$U_n(r_{a,b}) = [\kappa_{a,b}(n+1)] - [\kappa_{a,b}n],$$

where

$$\kappa_{a,b} = \frac{\log a}{\log(a/b)} \in (0, 1)$$

is not necessarily rational. Nevertheless $r_{a,b}$ assumes the same role as β in Section 10.1.

Let $\{p_n/q_n\}$ and $\{r_n/s_n\}$ be the lower and the upper approximation sequences associated with $\kappa_{a,b}$ respectively. Our induction is based on these sequences, unlike the previous section.

We introduce the proposition

$$\mathfrak{Q}\left(\kappa_{a,b}, \frac{r_n}{s_n}; \alpha, r_{a,b}\right)$$

defined by the condition:

(3) There exists $\alpha \in \{0\} \cup A^{\circ}_{s_n}$ satisfying

$$U_k(\alpha+) = U_k(r_{a,b})$$

for $1 \le k < s_{n+1} - 1$.

Recall that the first terms of the approximation sequences are $p_1/q_1 = 0/1$ and $r_1/s_1 = 1/1$. Furthermore

$$\frac{p_2}{q_2} = \frac{\ell_1 - 1}{\ell_1} \le \kappa_{a,b} < \frac{\ell_1}{\ell_1 + 1} = \frac{r_2}{s_2}$$

where

$$\ell_1 = \left[\frac{\kappa_{a,b}}{1 - \kappa_{a,b}}\right] + 1.$$

Hence it follows from Lemma 5.5 that

$$[\kappa_{a,b} k] = \left[\frac{\ell_1 - 1}{\ell_1} k\right]$$

for $1 \le k \le q_2 + s_2 - 1 = 2\ell_1$. Therefore we have

$$\begin{aligned} U_k(r_{a,b}) &= \left[\frac{\ell_1 - 1}{\ell_1}(k+1)\right] - \left[\frac{\ell_1 - 1}{\ell_1} k\right] \\ &= \left[\left\{\frac{\ell_1 - 1}{\ell_1} k\right\} + \frac{\ell_1 - 1}{\ell_1}\right] = 1 \end{aligned}$$

for $1 \le k < \ell_1 = s_2 - 1$, because $(\ell_1 - 1)k/\ell_1 \notin \mathbb{Z}$ if $\ell_1 \ge 2$. Thus the proposition

$$\mathfrak{Q}\left(\kappa_{a,b}, \frac{1}{1}; 0, r_{a,b}\right)$$

is certainly true.

Suppose now that $\mathfrak{Q}(\kappa_{a,b}, r_n/s_n; \alpha, r_{a,b})$ is true with $\alpha \in \{0\} \cup A^\circ_{s_n}$ for some integer $n \ge 1$. For simplicity we write p_n, q_n, r_n, s_n as p, q, r, s respectively. We also write $p_{n+1}, q_{n+1}, r_{n+1}, s_{n+1}$ as p', q', r', s' respectively. Then

$$\begin{pmatrix} p' \\ r' \end{pmatrix} = \begin{pmatrix} 1 & \ell - 1 \\ 1 & \ell \end{pmatrix} \begin{pmatrix} p \\ r \end{pmatrix} \quad \text{and} \quad \begin{pmatrix} q' \\ s' \end{pmatrix} = \begin{pmatrix} 1 & \ell - 1 \\ 1 & \ell \end{pmatrix} \begin{pmatrix} q \\ s \end{pmatrix}$$

hold, where

$$\ell = \left[\frac{\kappa_{a,b} q - p}{r - \kappa_{a,b} s}\right] + 1.$$

We go on considering step by step as follows:

1st Step: $u_{s'-1}(c)$ and $v_{s'-1}(c)$ are continuous on the interval $(\alpha, r_{a,b}]$.

The condition (3) and Lemma 9.8 imply that each function $u_k(c)$ is continuous and its graph does not get across the diagonal on the interval $(\alpha, r_{a,b})$ for $1 \le k < s' - 1$. Hence $u_{s'-1}(c)$ is continuous on $(\alpha, r_{a,b}]$. The same property holds for $v_k(c)$, because $U_k(\alpha+) = V_k(\alpha+)$ by Lemma 9.7 and $U_k(r_{a,b}) = V_k(r_{a,b})$ by (ii) of Lemma 9.5.

2nd Step: $\alpha < v_{s'-1}(\alpha+) < u_{s'-1}(\alpha+)$.

Since this is clear by (i) of Lemma 9.2 when $\alpha = 0$, we can assume that $\alpha \in A^\circ_s$ with $s \ge 2$. It follows from Lemma 9.7, the condition (3) and (ii) of Lemma 9.5

that

$$U_{s'-1}(\alpha+) = U_{q-1}(\alpha+)$$
$$= U_{q-1}(r_{a,b})$$
$$= [\kappa_{a,b}q] - [\kappa_{a,b}(q-1)].$$

Since $p/q \le \kappa_{a,b} < r/s$, we have from Lemma 5.5

$$U_{s'-1}(\alpha+) = \left[\frac{p}{q}q\right] - \left[\frac{p}{q}(q-1)\right] = 1,$$

because $q < q+s$, and therefore $V_{s'-1}(\alpha+) = 1$. By the same reason as in the second step in Section 10.1 we have $\alpha < v_{s'-1}(\alpha+) < u_{s'-1}(\alpha+)$.

3rd Step: $v_{s'-1}(r_{a,b}) = u_{s'-1}(r_{a,b}) < r_{a,b}$.

The equality follows from (iii) of Lemma 9.2. By (ii) of Lemma 9.5 we have

$$U_{s'-1}(r_{a,b}) = [\kappa_{a,b}s'] - [\kappa_{a,b}(s'-1)].$$

Since $p'/q' \le \kappa_{a,b} < r'/s'$, we have from Lemma 5.5

$$U_{s'-1}(r_{a,b}) = \left[\frac{p'}{q'}s'\right] - \left[\frac{p'}{q'}(s'-1)\right]$$
$$= \left[\frac{-1}{q'}\right] - \left[\frac{-1-p'}{q'}\right] = 0,$$

because $s' < q'+s'$, $p' < q'$ and $q'r' - p's' = 1$. We thus get $u_{s'-1}(r_{a,b}) < r_{a,b}$.

4th Step: From the considerations in the previous steps we see that the graphs of $u_{s'-1}(c)$ and $v_{s'-1}(c)$ intersect with the diagonal on the interval $(\alpha, r_{a,b})$. Let $\gamma > \delta$ be unique points in the interval $(\alpha, r_{a,b})$ satisfying $u_{s'-1}(\gamma) = \gamma$ and $v_{s'-1}(\delta) = \delta$. We denote the closed interval $[\delta, \gamma]$ by Δ. By the same argument as in the fourth step in Section 10.1 we have $\delta \in B_{s'}^{\circ}$ and $\gamma \in A_{s'}^{\circ}$.

5th Step: For any $c \in \Delta$ we have $\psi_{a,b,c} \in \mathfrak{B}_{s'}$.

This follows from exactly the same argument as in the fifth step in Section 10.1 by replacing β by $r_{a,b}$.

6th Step: $\rho(\psi_{a,b,c}) = \dfrac{r'}{s'}$ for any $c \in \Delta$.

As with the sixth step in Section 10.1 we put

$$L(c) = \sum_{k=1}^{s'} U_k(c) = \sum_{k=1}^{s'-1} U_k(c)$$

for any $c \in \varDelta$. We see that $U_k(c)$ is constant on the interval $(\alpha, r_{a,b}]$ for $1 \le k < s' - 1$ and $U_{s'-1}(c) = 1$. Then

$$\begin{aligned} L(c) &= \sum_{k=1}^{s'-2} U_k(r_{a,b}) + U_{s'-1}(c) \\ &= \sum_{k=1}^{s'-2} \big([\kappa_{a,b}(k+1)] - [\kappa_{a,b} k] \big) + 1 \\ &= [\kappa_{a,b}(s'-1)] + 1 = r'. \end{aligned}$$

7th Step: $U_k(\alpha+) = U_k(\delta)$ for $1 \le k < s + s' - 1$.

Since $U_k(c)$ is constant on the interval $(\alpha, r_{a,b}]$, we have $U_k(\alpha+) = U_k(\delta)$ for $1 \le k < s' - 1$. Moreover it has shown that $U_{s'-1}(\alpha+) = U_{s'-1}(\delta) = 1$. If $\alpha = 0$, then we have nothing to do any more, because $s = 1$. Suppose then that $\alpha \in A_s^\circ$.

We show the following equalities (i) and (ii):

(i) $U_q(r_{a,b}) = 0$.

Since $U_q(r_{a,b}) = [\kappa_{a,b}(q+1)] - [\kappa_{a,b} q]$ and $p/q \le \kappa_{a,b} < r/s$, we have

$$U_q(r_{a,b}) = \left[\frac{p}{q}(q+1) \right] - \left[\frac{p}{q} q \right] = 0,$$

because $q + 1 < q + s$.

(ii) $U_{k-s'}(r_{a,b}) = U_{k-s'+q}(r_{a,b})$ for $s' < k < s + s' - 1$.

Since $p/q \le \kappa_{a,b} < r/s$ and $k - s' + q + 1 < q + s$, we have from Lemma 5.5

$$\begin{aligned} U_{k-s'+q}(r_{a,b}) &= \left[\frac{p}{q}(k - s' + q + 1) \right] - \left[\frac{p}{q}(k - s' + q) \right] \\ &= \left[\frac{p}{q}(k - s' + 1) \right] - \left[\frac{p}{q}(k - s') \right] \\ &= U_{k-s'}(r_{a,b}). \end{aligned}$$

Replacing β by $r_{a,b}$ we can proceed to the same argument as in the seventh step in Section 10.1.

For $k = s' = q + \ell s$ we have

$$U_{s'}(\alpha+) = U_q(\alpha+) = U_q(r_{a,b}) = 0,$$

because of $q < s' - 1$, $s \ge 2$ and (i), while $U_{s'}(\delta) = 0$ clearly.

For $s' < k < s + s' - 1$ it follows from (ii) that

$$\begin{aligned} U_k(\alpha+) &= U_{k-\ell s}(\alpha+) \\ &= U_{k-\ell s}(r_{a,b}) \\ &= U_{k-s'}(r_{a,b}) = U_{k-s'}(\alpha+), \end{aligned}$$

because $k - \ell s = k - s' + q < q + s - 1$ and $k - s' < s - 1 < s' - 1$. On the other hand, we have

$$U_k(\delta) = U_{k-s'}(\delta) = U_{k-s'}(\alpha+).$$

Last Step: $U_k(\gamma+) = U_k(r_{a,b})$ for $1 \le k < s'' - 1$, where $s'' = s_{n+2}$.

Since $r'/s' < 1$, it follows from Theorem 5.1 that the first $s'' - 2$ digits of the sequence

$$\mathfrak{S}(\kappa_{a,b}) = \{U_n(r_{a,b})\}_{n\ge1}$$

coincide with that of

$$\mathfrak{S}^*(r'/s') = \{U_n(\gamma+)\}_{n\ge1}.$$

The above considerations imply that both the propositions

$$\mathfrak{P}\left(\frac{r'}{s'}, \frac{r}{s}; \alpha, \delta\right) \quad \text{and} \quad \mathfrak{Q}\left(\kappa_{a,b}, \frac{r'}{s'}; \gamma, r_{a,b}\right)$$

are true. Thus we can apply the induction argument in the previous section to the former and that in this section to the latter respectively. Repeating these arguments infinitely often, we therefore have, in a similar way with the previous section,

Theorem 10.2. *Suppose that* $0 < b < 1 < a$. *For each irreducible fraction* p/q *in the interval* $(\kappa_{a,b}, 1)$ *there exists a closed interval* $\Delta(p/q) \subset (0, r_{a,b})$ *satisfying the following properties:*

(i) *If* $p/q < r/s$, *then* $\Delta(p/q) \succ \Delta(r/s)$.
(ii) *According as c is the left endpoint, an interior point or the right endpoint of* $\Delta(p/q)$, *the piecewise linear function* $\psi_{a,b,c}$ *belongs to*

$$\mathfrak{E}^{\phi}\left(\frac{p}{q}\right), \quad \mathfrak{E}^{*}\left(\frac{p}{q}\right) \quad \text{or} \quad \mathfrak{E}^{0}\left(\frac{p}{q}\right).$$

10.3 Residual Sets

For arbitrarily fixed parameters a, b satisfying $a \geq b$ and $0 < b < 1$ let $J_{a,b}$ be the open interval defined by

$$J_{a,b} = \begin{cases} (0,1) & (a \leq 1), \\ (\kappa_{a,b}, 1) & (a > 1). \end{cases}$$

We then put

$$\Gamma_{a,b} = [0, r_{a,b}] \setminus \bigcup_{p/q \in J_{a,b}} \Delta\left(\frac{p}{q}\right),$$

where the union in the right-hand side extends over all irreducible fractions p/q in the interval $J_{a,b}$. The set $\Gamma_{a,b}$ is the remainder obtained from the interval $[0, r_{a,b}]$ by an infinite sequence of deletions of disjoint closed intervals. It is not hard to see that $\Gamma_{a,b}$ is uncountable. We call $\Gamma_{a,b}$ the *residual set* and this terminology might be justified by the following

Theorem 10.3. *For any $c \in \Gamma_{a,b}$ we have $\psi_{a,b,c} \in \mathfrak{B}_\infty$.*

Proof. For all sufficiently large integer n we can take two consecutive Farey fractions $a_n/b_n < c_n/d_n$ in $\mathfrak{F}_n \setminus \{0/1, 1/1\}$ satisfying ‡

$$\Delta\left(\frac{c_n}{d_n}\right) < c < \Delta\left(\frac{a_n}{b_n}\right)$$

where $\Delta(c_n/d_n)$ is the rightmost interval less than c among the intervals $\Delta(p/q)$, $p/q \in \mathfrak{F}_n$, and $\Delta(a_n/b_n)$ is the leftmost one greater than c. Put $\Delta(c_n/d_n) = [\alpha_n, \beta_n]$ and $\Delta(a_n/b_n) = [\gamma_n, \delta_n]$. Then $\beta_n < c < \gamma_n$.

By the definition of $\Gamma_{a,b}$ either b_n or d_n must tend to ∞ as $n \to \infty$. Since the proposition

$$\mathfrak{P}\left(\frac{a_n}{b_n}, \frac{c_n}{d_n}; \beta_n, \gamma_n\right)$$

holds, it follows from the condition (2) in Section 10.1 and Lemmas 9.6, 9.7 that $\psi^k(1) < \psi^k(0)$ and $\epsilon \circ \psi^k(0) = \epsilon \circ \psi^k(1)$ for $1 \leq k < b_n + d_n - 1$, where $\psi = \psi_{a,b,c}$. This means that $K_k = [\psi^k(1), \psi^k(0))$ is defined and $c \notin \overline{K_k}$ for all $k \geq 1$, because n can be taken to be arbitrarily large. Hence ψ belongs to $\mathfrak{B}_\infty$. □

‡ We identify the number c with the singleton set having exactly one point c. Note that c_n is a positive integer, unrelated to any image of c.

10.4 Length of $\Delta(p/q)$

In this section we calculate the length of the interval $\Delta(p/q)$ stated in Theorems 10.1 and 10.2 for any irreducible fraction p/q in the interval $J_{a,b}$.

We put $\Delta(p/q) = [\beta, \alpha]$ with $\alpha \in A_q^\circ$ and $\beta \in B_q^\circ$. As is already seen in the proof of Lemma 9.6, the function $\psi = \psi_{a,b,\beta}$ belongs to $\mathfrak{E}^\phi(p/q)$. Therefore it follows from the result in the proof of Theorem 8.1 that

$$\begin{aligned} u_n(\beta) - v_n(\beta) &= \psi^n(0) - \psi^n(1) = |K_n| \\ &= (1 - b - (a-b)\beta)a^{n-1}\left(\frac{b}{a}\right)^{[pn/q]} \end{aligned} \tag{10.1}$$

for all $n \geq 1$.

Each graph of the orbital function $u_n(c)$ does not pass transversely across the diagonal on $\Delta(p/q)$ by Theorem 9.1. Hence $u_n(c)$ is a linear function with slope, say $-\sigma_n$, on this interval. For example, we have $\sigma_1 = a$, because $u_1(c) = 1 - ac$.

It follows from the recurrence formula (9.1) that

$$\sigma_{n+1} = \begin{cases} a(1+\sigma_n) & (U_n(\beta) = 0), \\ b(1+\sigma_n) & (U_n(\beta) = 1), \end{cases}$$

which yields that

$$\sigma_{n+1} = a\left(\frac{b}{a}\right)^{[p(n+1)/q]-[pn/q]}(1+\sigma_n),$$

because $U_n(\beta) = [p(n+1)/q] - [pn/q]$. Therefore

$$\sigma_n = \sum_{k=1}^{n} a^k\left(\frac{b}{a}\right)^{[pn/q]-[p(n-k)/q]}. \tag{10.2}$$

Since $u_{q-1}(\alpha) = \alpha$ and $v_{q-1}(\beta) = \beta$, it can be easily seen that

$$\left|\Delta\left(\frac{p}{q}\right)\right| = \alpha - \beta = \frac{u_{q-1}(\beta) - v_{q-1}(\beta)}{1 + \sigma_{q-1}}.$$

So it follows from (10.1) that

$$u_{q-1}(\beta) - v_{q-1}(\beta) = (1 - b - (a-b)\beta)a^{q-p-1}b^{p-1}$$

and from (10.2) that

$$\begin{aligned} 1 + \sigma_{q-1} &= 1 + \sum_{k=1}^{q-1} a^k\left(\frac{b}{a}\right)^{-1-[-p(k+1)/q]} \\ &= P_{p,q}^+\left(a, \frac{b}{a}\right). \end{aligned}$$

Using the basic formula of the first kind (Theorem 8.1) we thus have the following

Theorem 10.4. *For any fixed a, b with $a \geq b$, $0 < b < 1$ and for any irreducible fraction p/q in the interval $J_{a,b}$ we have*

$$\left|\Delta\left(\frac{p}{q}\right)\right| = \frac{a^{q-p-1}b^{p-1}(1 - a^{q-p}b^{p})}{P_{p,q}(a, b/a)\,P^{+}_{p,q}(a, b/a)}.$$

10.5 Distance between Two Δ's

We look anew at the interval $\Delta(p/q) = [\beta, \alpha]$, writing it as $[\beta(p/q), \alpha(p/q)]$. Since $\psi_{a,b,\alpha} \in \mathfrak{E}^0(p/q)$, Theorems 8.3 and 8.4 imply that

$$\alpha\left(\frac{p}{q}\right) = 1 - \frac{1}{a} \cdot \frac{P_{q,p}(b/a, a)}{P^{+}_{p,q}(a, b/a)} \tag{10.3}$$

for any irreducible fraction p/q in the interval $(0, 1)$. The right-hand side of (10.3) has still a meaning even if $p/q = 1/1$; so we define $\alpha(1/1) = 0$. On the other hand, since

$$\begin{aligned}\alpha\left(\frac{1}{q}\right) &= 1 - \frac{1}{a} \cdot \frac{P_{q,1}(b/a, a)}{P^{+}_{1,q}(a, b/a)} \\ &= 1 - \frac{a^{q-1}}{1 + a + \cdots + a^{q-1}},\end{aligned}$$

we see that the sequence $\alpha(1/q)$ tends to 1 as $q \to \infty$ if $a \leq 1$; in other words, $c = 1$ is an accumulation point of Δ's when $r_{a,b} = 1$. The similar property for the case $r_{a,b} < 1$ will be discussed in the next section.

Since $\psi_{a,b,\beta} \in \mathfrak{E}^{\phi}(p/q)$, Theorems 8.1 and 8.2 imply that

$$\beta\left(\frac{p}{q}\right) = 1 - \frac{1}{a} \cdot \frac{P^{+}_{q,p}(b/a, a)}{P_{p,q}(a, b/a)} \tag{10.4}$$

for any irreducible fraction p/q in the interval $(0, 1)$. By adopting $P^{+}_{1,0}(z, w) = 0$ as an empty sum, we define $\beta(0/1) = 1$. On the other hand, since

$$\begin{aligned}\beta\left(\frac{q-1}{q}\right) &= 1 - \frac{1}{a} \cdot \frac{P^{+}_{q,q-1}(b/a, a)}{P_{q-1,q}(a, b/a)} \\ &= \frac{b^{q-1}}{1 + b + \cdots + b^{q-1}},\end{aligned}$$

we see that the sequence $\beta(1 - 1/q)$ tends to 0 as $q \to \infty$ because $0 < b < 1$; in other words, $c = 0$ is always an accumulation point of Δ's.

We combine (10.3), (10.4) and Theorem 10.4 to make the following

Lemma 10.1. *For any irreducible fraction p/q in the interval $(0, 1)$ we have the identity*:

$$P_{p,q}(w,z)\,P_{q,p}(z,w) - P^{+}_{q,p}(z,w)P^{+}_{p,q}(w,z) = w^{q-1}z^{p-1}(w^{q}z^{p}-1). \qquad (10.5)$$

Note that (10.5) holds for the case $p/q = 1/1$, because $P^{+}_{1,1}(z,w) = 1$ and $P_{1,1}(w,z) = z$. Furthermore, to make (10.5) valid even if $p/q = 0/1$, we must exceptionally define

$$P_{1,0}(z,w) = \frac{w-1}{z},$$

not as an empty sum. This might look strange, but it enables us to make induction arguments so simple. Thus (10.5) holds for any irreducible fraction p/q in the interval $[0, 1]$.

To investigate the residual set $\Gamma_{a,b}$ in detail we need to calculate the distance between $\Delta(p/q)$ and $\Delta(r/s)$:

$$\operatorname{dist}\left(\frac{p}{q}, \frac{r}{s}\right) = \beta\left(\frac{p}{q}\right) - \alpha\left(\frac{r}{s}\right).$$

Lemma 10.2. *For any fractions p/q and r/s in the interval $[0, 1]$ satisfying $qr - ps = 1$ we have the identity*:

$$P_{p,q}(w,z)\,P_{s,r}(z,w) - P^{+}_{q,p}(z,w)P^{+}_{r,s}(w,z) = w^{q+s-1}z^{p+r-1}. \qquad (10.6)$$

Proof. Let $\mathfrak{P}(p/q, r/s)$ be the proposition that (10.6) holds identically. We first show the case $p/q = 0/1$; so, $r = 1$ and $s \geq 1$. Since

$$P_{0,1}(w,z) = 1, \quad P_{s,1}(z,w) = w^{s} \quad \text{and} \quad P^{+}_{1,0}(z,w) = 0,$$

the proposition $\mathfrak{P}(0/1, 1/s)$ certainly holds for all $s \geq 1$.

We next assume that the proposition $\mathfrak{P}(p/q, r/s)$ holds for some fractions p/q and r/s in the interval $[0, 1]$ satisfying $qr - ps = 1$.

(a) We first deal with the proposition $\mathfrak{P}(p/q, (p+r)/(q+s))$. From the previous consideration we can assume that $p \geq 1$. Applying Lemma 5.8 to s/r and q/p, the formula (5.1) implies that

$$P_{q+s,p+r}(z,w) = P_{s,r}(z,w) + w^{s}z^{r}P_{q,p}(z,w). \qquad (10.7)$$

Similarly, applying Lemma 5.8 to p/q and r/s, the formula (5.2) implies that

$$P^{+}_{p+r,q+s}(w,z) = P^{+}_{r,s}(w,z) + w^{s}z^{r}P^{+}_{p,q}(w,z). \qquad (10.8)$$

Substituting the formulae (10.7) and (10.8) into the expression

$$\Phi = P_{p,q}(w,z)\,P_{q+s,p+r}(z,w) - P^{+}_{q,p}(z,w)P^{+}_{p+r,q+s}(w,z),$$

we obtain $\Phi = \Phi_0 + w^s z^r \Phi_1$ where

$$\Phi_0 = P_{p,q}(w,z)\,P_{s,r}(z,w) - P^+_{q,p}(z,w)\,P^+_{r,s}(w,z)$$

and

$$\Phi_1 = P_{p,q}(w,z)\,P_{q,p}(z,w) - P^+_{q,p}(z,w)P^+_{p,q}(w,z).$$

Since $\mathfrak{P}(p/q, r/s)$ is true, it follows from Lemma 10.1 that

$$\Phi_0 = w^{q+s-1}z^{p+r-1} \quad \text{and} \quad \Phi_1 = w^{q-1}z^{p-1}(w^q z^p - 1).$$

We thus obtain

$$\Phi = w^{2q+s-1}z^{2p+r-1},$$

which means that the proposition $\mathfrak{P}(p/q, (p+r)/(q+s))$ is also true.

(b) We next deal with the proposition $\mathfrak{P}((p+r)/(q+s), r/s)$. Applying Lemma 5.8 to p/q and r/s, the formula (5.1) implies that

$$P_{p+r,q+s}(w,z) = P_{p,q}(w,z) + w^q z^p P_{r,s}(w,z). \tag{10.9}$$

Similarly, applying Lemma 5.8 to s/r and q/p, the formula (5.2) implies that

$$P^+_{q+s,p+r}(z,w) = P^+_{q,p}(z,w) + w^q z^p P^+_{s,r}(z,w). \tag{10.10}$$

Substituting the formulae (10.9) and (10.10) into the expression

$$G = P_{p+r,q+s}(w,z)\,P_{s,r}(z,w) - P^+_{q+s,p+r}(z,w)P^+_{r,s}(w,z),$$

we get $G = G_0 + w^q z^p G_1$ where

$$G_0 = P_{p,q}(w,z)P_{s,r}(z,w) - P^+_{q,p}(z,w)\,P^+_{r,s}(w,z) = \Phi_0$$

and

$$G_1 = P_{r,s}(w,z)\,P_{s,r}(z,w) - P^+_{s,r}(z,w)P^+_{r,s}(w,z).$$

Therefore, since $\mathfrak{P}(p/q, r/s)$ is true, it follows from Lemma 10.1 that

$$G_0 = w^{q+s-1}z^{p+r-1} \quad \text{and} \quad G_1 = w^{s-1}z^{r-1}(w^s z^r - 1).$$

We thus get

$$G = w^{q+2s-1}z^{p+2r-1},$$

which means that the proposition $\mathfrak{P}((p+r)/(q+s), r/s)$ is also true.

Hence, by Lemma 5.4 we conclude that the proposition $\mathfrak{P}(p/q, r/s)$ is true for all fractions p/q and r/s in the interval $[0,1]$ satisfying $qr - ps = 1$. □

As a straightforward application of the above lemma, we have

Theorem 10.5. *For any fractions p/q, r/s in the interval $J_{a,b}$ satisfying $qr-ps=1$ we have*

$$\mathrm{dist}\left(\frac{p}{q}, \frac{r}{s}\right) = \frac{a^{q+s-p-r-1} b^{p+r-1}}{P_{p,q}(a, b/a) P_{r,s}^{+}(a, b/a)}.$$

Moreover, if $J_{a,b} = (0, 1)$, then it can be replaced by its closure $\overline{J}_{a,b} = [0, 1]$.

10.6 Some Properties when $r_{a,b} < 1$

In the previous section we have seen that $c = 1$ is an accumulation point of Δ's when $r_{a,b} = 1$. The similar property holds even if $r_{a,b} < 1$. To see this, we first need the following lemma:

Lemma 10.3. *For any fractions p/q, r/s in the interval $[0, 1]$ satisfying $qr-ps=1$ and $q > s$, we have*

$$P_{p,q}(w, z) > P_{r,s}(w, z) \quad \text{and} \quad P_{p,q}^{+}(w, z) > P_{r,s}^{+}(w, z)$$

for any $w > 0$ and $0 < z < 1$.

Proof. We have $p \geq 1$, because $q \geq 2$. The fractions p/q and r/s belong to the same Farey interval of order $s-1$. For otherwise, there exists a fraction u/v satisfying $p/q < u/v \leq r/s$ and $v < s$. Since

$$\frac{1}{qs} \geq \frac{qu - pv}{qv} \geq \frac{1}{qv},$$

we have $v \geq s$, a contradiction. Hence it follows from Lemma 5.5 that $[pk/q] = [rk/s]$ for $1 \leq k < s$. Thus

$$P_{p,q}(w, z) - P_{r,s}(w, z) \geq w^{s-1} z^{r-1}(1 - z) > 0.$$

On the other hand, $\lceil pk/q \rceil = \lceil rk/s \rceil$ holds for $1 \leq k \leq s$, and hence $P_{p,q}^{+}(w, z) > P_{r,s}^{+}(w, z)$. This completes the proof. □

Theorem 10.6. *For any fixed $0 < b < 1 < a$, $r_{a,b}$ is an accumulation point of Δ's.*

Proof. Suppose, on the contrary, that $r_{a,b}$ is not an accumulation point of Δ's. This means that the limit of the strictly monotonically increasing sequence $\{\alpha(r_n/s_n)\}$, say α^*, is strictly less than $r_{a,b}$, where $\{r_n/s_n\}$ is the upper approximation sequence associated with $\kappa_{a,b}$. Since $\psi_{a,b,\alpha(r_n/s_n)} \in \mathfrak{E}^0(r_n/s_n)$, it follows from the basic formula of the first kind (Theorem 8.3) that

$$r_{a,b} - \alpha\left(\frac{r_n}{s_n}\right) = \frac{1 - b - (a-b)\alpha(r_n/s_n)}{a-b}$$
$$= \frac{1}{a-b} \cdot \frac{1 - a^{s_n - r_n} b^{r_n}}{P^+_{r_n,s_n}(a, b/a)}. \tag{10.11}$$

The sequence $\{P^+_{r_n,s_n}(a, b/a)\}$ converges to a finite limit or diverges to infinity as $n \to \infty$ by Lemma 10.3. However, if it diverges, (10.11) would imply $\alpha^* = r_{a,b}$, contrary to the assumption. Therefore $P^+_{r_n,s_n}(a, b/a)$ converges to a finite limit. So, the sequence $\{a^{s_n - r_n} b^{r_n}\}$ also converges to some $\lambda^* \in [0, 1)$ as $n \to \infty$. Similarly, we have

$$r_{a,b} - \beta\left(\frac{r_n}{s_n}\right) = \frac{1}{a-b} \cdot \frac{1 - a^{s_n - r_n} b^{r_n}}{P_{r_n,s_n}(a, b/a)}$$

from Theorem 8.1, which implies that $P_{r_n,s_n}(a, b/a)$ also converges to a finite limit. Since $|\Delta(r_n/s_n)| \to 0$ as $n \to \infty$, it follows from Theorem 10.4 that $\lambda^* = 0$, and hence there exists some integer m satisfying

$$a^{s_m - r_m} b^{r_m} < \frac{b}{a}.$$

Therefore, using the relation $a = (a/b)^{\kappa_{a,b}}$ we have

$$\left(\frac{b}{a}\right)^{r_m - \kappa_{a,b} s_m} < \frac{b}{a},$$

which yields $r_m - \kappa_{a,b} s_m > 1$. However this is a contradiction, because

$$1 < r_m - \kappa_{a,b} s_m$$
$$\le s_m\left(\frac{r_m}{s_m} - \frac{p_m}{q_m}\right) = \frac{1}{q_m},$$

where $\{p_n/q_n\}$ is the lower approximation sequence for $\kappa_{a,b}$. This completes the proof. □

We finally remark that, if $\kappa_{a,b}$ is a rational number, say u/v, then the equalities

$$\alpha\left(\frac{u}{v}\right) = \beta\left(\frac{u}{v}\right) = r_{a,b}$$

hold. To see this, using $b = a^{1-v/u}$, we have

$$P_{v,u}\left(\frac{b}{a}, a\right) = \sum_{k=1}^{u} a^{v/u-\{vk/u\}}$$

$$= \sum_{i=0}^{u-1} a^{v/u-i/u}$$

$$= \frac{a^{v/u} - a^{v/u-1}}{1 - a^{-1/u}}$$

and

$$P_{u,v}^{+}\left(a, \frac{b}{a}\right) = \sum_{k=1}^{v} a^{v/u-1-v/u(\lceil uk/v \rceil - uk/v)}$$

$$= \sum_{i=0}^{v-1} a^{v/u-1-i/u}$$

$$= \frac{a^{v/u-1} - a^{-1}}{1 - a^{-1/u}},$$

and hence from (10.3)

$$\alpha\left(\frac{u}{v}\right) = \frac{a^{v/u-1} - 1}{a^{v/u} - 1} = \frac{1 - b}{a - b}.$$

Similarly we obtain

$$P_{u,v}\left(a, \frac{b}{a}\right) = \sum_{k=1}^{v} a^{v/u\{uk/v\}-1} = \frac{a^{v/u-1} - a^{-1}}{a^{1/u} - 1}$$

and

$$P_{v,u}^{+}\left(\frac{b}{a}, a\right) = \sum_{k=1}^{u} a^{v/u-1+\lceil vk/u \rceil - vk/u} = \frac{a^{v/u} - a^{v/u-1}}{a^{1/u} - 1}.$$

Thus we get from (10.4)

$$\beta\left(\frac{u}{v}\right) = \frac{a^{v/u-1} - 1}{a^{v/u} - 1} = \frac{1 - b}{a - b},$$

as required.

10.7 Proof of $|\Gamma_{a,b}| = 0$ when $a < 1$

In this section we will show that the residual set $\Gamma_{a,b}$ is a null set when $0 < b \leq a < 1$. For any fractions p/q and r/s in the interval $J_{a,b}$ satisfying $qr-ps = 1$ we denote by $\varrho(p/q, r/s)$ the ratio of the length of the intermediate interval $\Delta((p+r)/(q+s))$ to the distance between two intervals $\Delta(p/q)$ and $\Delta(r/s)$; that is,

$$\varrho\left(\frac{p}{q}, \frac{r}{s}\right) = \frac{\left|\Delta\left(\frac{p+r}{q+s}\right)\right|}{\operatorname{dist}\left(\frac{p}{q}, \frac{r}{s}\right)}.$$

If $J_{a,b} = (0, 1)$, then we allow the cases $p/q = 0/1$ or $r/s = 1/1$, because "dist" is well-defined for these values. Then from Theorems 10.4 and 10.5 we have

$$\varrho\left(\frac{p}{q}, \frac{r}{s}\right) = \frac{P_{p,q}(a, b/a)\, P^+_{r,s}(a, b/a)}{P_{p+r,q+s}(a, b/a)\, P^+_{p+r,q+s}(a, b/a)}(1 - a^{q+s-p-r}b^{p+r}). \tag{10.12}$$

Put $\lambda_0 = \min(1 - a, b/a)$ for brevity.

For any fraction p/q in the interval $[0, 1]$ it follows plainly that

$$\begin{aligned} P_{p,q}\left(a, \frac{b}{a}\right) &\geq \left(\frac{b}{a}\right)^{[p/q]} \\ &\geq \frac{b}{a} \geq \lambda_0 \end{aligned}$$

and

$$\begin{aligned} P_{p,q}\left(a, \frac{b}{a}\right) &< \sum_{k=1}^{\infty} a^{k-1}\left(\frac{b}{a}\right)^{[pk/q]} \\ &\leq \frac{1}{1-a} \leq \frac{1}{\lambda_0}. \end{aligned}$$

Since these lower and upper estimates are valid for $P^+_{p,q}(a, b/a)$, it follows from (10.12) that

$$\begin{aligned} \varrho\left(\frac{p}{q}, \frac{r}{s}\right) &\geq (1 - a^2)\lambda_0^4 \\ &\geq \lambda_0^5 = \lambda_1, \end{aligned}$$

and hence

$$\left|\Delta\left(\frac{p+r}{q+s}\right)\right| \geq \lambda_1 \operatorname{dist}\left(\frac{p}{q}, \frac{r}{s}\right). \tag{10.13}$$

Let $\mathcal{G}_n$ be the series defined in Exercise 3 in Chapter 5, constructed from two fractions $0/1$ and $1/1$ without any condition on denominators. Then $\cup_{n=1}^{\infty} \mathcal{G}_n$ is the set of all irreducible fractions in the interval $[0, 1]$. Here are some examples:

$$\mathcal{G}_1 = \left\{\frac{0}{1}, \frac{1}{1}\right\},$$
$$\mathcal{G}_2 = \left\{\frac{0}{1}, \frac{1}{2}, \frac{1}{1}\right\},$$
$$\mathcal{G}_3 = \left\{\frac{0}{1}, \frac{1}{3}, \frac{1}{2}, \frac{2}{3}, \frac{1}{1}\right\},$$
$$\mathcal{G}_4 = \left\{\frac{0}{1}, \frac{1}{4}, \frac{1}{3}, \frac{2}{5}, \frac{1}{2}, \frac{3}{5}, \frac{2}{3}, \frac{3}{4}, \frac{1}{1}\right\}.$$

We also put $\mathcal{G}_n^\circ = \mathcal{G}_n \setminus \{0/1, 1/1\}$ for all $n \geq 2$.

Let $\mathfrak{P}_n$ be the proposition that the estimate

$$\left| [0, 1] \setminus \bigcup_{p/q \in \mathcal{G}_n^\circ} \Delta\left(\frac{p}{q}\right) \right| \leq (1 - \lambda_1)^{n-1}$$

holds. Obviously $\mathfrak{P}_2$ holds, because $|\Delta(1/2)| \geq \lambda_1$.

Suppose next that the proposition $\mathfrak{P}_n$ holds for some $n \geq 2$. For any consecutive fractions $p/q < r/s$ in $\mathcal{G}_n$ the mediant $(p + r)/(q + s)$ belongs to $\mathcal{G}_{n+1}$ and the intermediate interval $\Delta(p + r)/(q + s)$ lies between $\Delta(p/q)$ and $\Delta(r/s)$. Therefore it follows from (10.13) that

$$\begin{aligned}
\left| [0, 1] \setminus \bigcup_{p/q \in \mathcal{G}_{n+1}^\circ} \Delta\left(\frac{p}{q}\right) \right| &= \sum_{\substack{p/q, r/s \in \mathcal{G}_{n+1} \\ qr - ps = 1}} \operatorname{dist}\left(\frac{p}{q}, \frac{r}{s}\right) \\
&= \sum_{\substack{p/q, r/s \in \mathcal{G}_n \\ qr - ps = 1}} \left(\operatorname{dist}\left(\frac{p}{q}, \frac{r}{s}\right) - \left| \Delta\left(\frac{p + r}{q + s}\right) \right| \right) \\
&\leq (1 - \lambda_1) \sum_{\substack{p/q, r/s \in \mathcal{G}_n \\ qr - ps = 1}} \operatorname{dist}\left(\frac{p}{q}, \frac{r}{s}\right) \\
&\leq (1 - \lambda_1)^n,
\end{aligned}$$

and hence the proposition $\mathfrak{P}_{n+1}$ also holds. Therefore

$$\left| \Gamma_{a,b} \right| \leq (1 - \lambda_1)^n$$

for all $n \geq 2$. We thus have the following

Theorem 10.7. *For any $0 < b \leq a < 1$ we have $|\Gamma_{a,b}| = 0$.*

10.8 Proof of $|\Gamma_{a,b}| = 0$ when $a \geq 1$

We next show that the residual set $\Gamma_{a,b}$ is a null set for any $0 < b < 1 \leq a$. Let $\{r_n/s_n\}$ be the upper approximation sequence associated with $\kappa_{a,b}$. For any integer $m \geq 2$ we consider the closed subinterval

$$T_m = \left[\alpha\left(\frac{r_{m-1}}{s_{m-1}}\right), \beta\left(\frac{r_m}{s_m}\right)\right]$$

of $[0, r_{a,b}]$. Then, for any irreducible fraction p/q in the interval

$$W_m = \left[\frac{r_m}{s_m}, \frac{r_{m-1}}{s_{m-1}}\right],$$

we have

$$a\left(\frac{b}{a}\right)^{p/q} \leq a\left(\frac{b}{a}\right)^{r_m/s_m} = \xi_m < 1,$$

because $\kappa_{a,b} < r_m/s_m$ and $a(b/a)^{\kappa_{a,b}} = 1$. Thus,

$$\begin{aligned} P_{p,q}\left(a, \frac{b}{a}\right) &< \sum_{k=1}^{\infty} a^{k-1}\left(\frac{b}{a}\right)^{[pk/q]} \\ &\leq \frac{1}{b}\sum_{k=1}^{\infty} \xi_m^k \\ &< \frac{1}{b} \cdot \frac{1}{1-\xi_m}. \end{aligned}$$

Therefore, replacing λ_0 by

$$\lambda_2 = \min\left(\frac{b}{a}, b(1-\xi_m)\right),$$

we can apply the same argument as in Section 10.7 to this case so that we have

$$\varrho\left(\frac{p}{q}, \frac{r}{s}\right) \geq \lambda_2^5 = \lambda_3$$

for any fractions p/q and r/s in the interval W_m satisfying $qr - ps = 1$.

Let $\mathcal{G}_n^m$ be the series defined in Exercise 3 in Chapter 5, constructed from two fractions r_m/s_m and r_{m-1}/s_{m-1} without any condition on denominators. Then $\cup_{n=1}^{\infty} \mathcal{G}_n^m$ is the set of all irreducible fractions in the closed interval W_m. Applying then a similar argument as in Section 10.7 to the series $\mathcal{G}_n^m$, we obtain

$$\left|\Gamma_{a,b} \cap T_m\right| = 0.$$

Therefore we have $|\Gamma_{a,b}| = 0$, because

$$[0, r_{a,b}) = \bigcup_{m=2}^{\infty} \left(T_m \cup \Delta\left(\frac{r_m}{s_m}\right)\right)$$

from Theorem 10.6. Thus,

Theorem 10.8. *For any* $0 < b < 1 \le a$ *we have* $|\Gamma_{a,b}| = 0$.

10.9 Summation Formulae

We will discuss in this section some interesting formulae derived from Theorems 10.7 and 10.8. The following makes a paraphrase of these theorems.

Theorem 10.9. *For any* $0 < b < 1$, $a \ge b$ *we have*

$$\sum_{p/q \in J_{a,b}} \frac{a^{q-p-1} b^{p-1}(1 - a^{q-p} b^p)}{P_{p,q}(a, b/a) P^{+}_{p,q}(a, b/a)} = r_{a,b},$$

where the sum extends over all irreducible fractions p/q *in the open interval* $J_{a,b}$.

In particular, when $0 < a = b < 1$, we have

$$\sum_{n=2}^{\infty} \varphi(n) \frac{a^n}{1 - a^n} = \left(\frac{a}{1 - a}\right)^2, \tag{10.14}$$

where $\varphi(n)$ is Euler's totient function. This form of series is known as a *Lambert series*.

As another example, we consider a rearrangement of the sum in Theorem 10.9 when $a = 1$. For any fraction p/q with $p \ge 2$ in the interval $(0, 1)$ it is irreducible if and only if $q = \ell p + d$ for $\ell \ge 1$ and $1 \le d < p$ with $\gcd(d, p) = 1$. When $p = 1$, we write $q = \ell + 1$, $\ell \ge 1$. Thus, for any $0 < b < 1$ we have

$$S_{1,1}(b) + \sum_{\substack{p \ge 2, 1 \le d < p \\ \gcd(d,p)=1}} S_{d,p}(b) = 1,$$

where

$$S_{d,p}(b) = \sum_{\ell=1}^{\infty} \frac{b^{p-1}(1 - b^p)}{P_{p,\ell p+d}(1, b) P^{+}_{p,\ell p+d}(1, b)}.$$

We now need the following

Lemma 10.4. *For any $\ell \geq 0$, $p \geq 2$, $1 \leq d < p$ with* $\gcd(d, p) = 1$ *or* $p = d = 1$, *we have*

$$P_{p,\ell p+d}(1, z) = \ell \frac{1 - z^p}{1 - z} + P_{p,d}(1, z)$$

and

$$P^+_{p,\ell p+d}(1, z) = \ell \frac{1 - z^p}{1 - z} + P^+_{p,d}(1, z).$$

Proof. We first suppose that $p \geq 2$ and $1 \leq d < p$ with $\gcd(d, p) = 1$. Put

$$P_{p,\ell p+d}(1, z) = \sum_{k=0}^{p} a_k^\ell z^k,$$

where a_k^ℓ is the number of integral solutions in j of $[pj/(\ell p + d)] = k$ in the closed interval $[1, \ell p + d]$. Equivalently, a_k^ℓ is the number of $j \in [1, \ell p + d]$ satisfying

$$\left(\ell + \frac{d}{p}\right)k \leq j < \left(\ell + \frac{d}{p}\right)(k + 1).$$

Hence we get

$$\begin{aligned} a_k^\ell &= \left\lceil \left(\ell + \frac{d}{p}\right)(k + 1) \right\rceil - \left\lceil \left(\ell + \frac{d}{p}\right)k \right\rceil \\ &= \ell + \left\lceil \frac{d(k + 1)}{p} \right\rceil - \left\lceil \frac{dk}{p} \right\rceil \\ &= \ell + a_k^0 \end{aligned}$$

for $0 < k < p$, $a_0^\ell = \ell$ and $a_p^\ell = 1$. Therefore we obtain

$$P_{p,\ell p+d}(1, z) - P_{p,d}(1, z) = \ell(1 + z + \cdots + z^{p-1}).$$

Since $P_{1,\ell+1}(1, z) = \ell + z$, this is true even if $d = p = 1$.

Similarly we put

$$P^+_{p,\ell p+d}(1, z) = \sum_{k=0}^{p-1} b_k^\ell z^k,$$

where b_k^ℓ is the number of $j \in [1, \ell p + d]$ satisfying

$$\left(\ell + \frac{d}{p}\right)k < j \leq \left(\ell + \frac{d}{p}\right)(k + 1).$$

Then we have

$$\begin{aligned} b_k^\ell &= \left\lfloor \left(\ell + \frac{d}{p}\right)(k + 1) \right\rfloor - \left\lfloor \left(\ell + \frac{d}{p}\right)k \right\rfloor \\ &= \ell + b_k^0 \end{aligned}$$

for $0 \leq k < p$. Therefore

$$P^+_{p,\ell p+d}(1,z) - P^+_{p,d}(1,z) = \ell(1 + z + \cdots + z^{p-1}).$$

Since $P^+_{1,\ell+1}(1,z) = \ell + 1$, this is true if $d = p = 1$. This completes the proof. □

Using the relation

$$P^+_{p,d}(1,b) = P_{p,d}(1,b) + b^{p-1}(1-b)$$

and Lemma 10.4 the series $S_{d,p}(b)$ can be written as

$$\frac{b^{p-1}(1-b)^2}{1-b^p}\sum_{\ell=1}^{\infty}\left(\ell + \frac{P_{p,d}(1,b)}{1+b+\cdots+b^{p-1}}\right)^{-1}\left(\ell + \frac{P^+_{p,d}(1,b)}{1+b+\cdots+b^{p-1}}\right)^{-1}$$

$$= \sum_{\ell=1}^{\infty}\left(\left(\ell + \frac{P_{p,d}(1,b)}{1+b+\cdots+b^{p-1}}\right)^{-1} - \left(\ell + \frac{P^+_{p,d}(1,b)}{1+b+\cdots+b^{p-1}}\right)^{-1}\right).$$

Let $\Gamma(z)$ be the gamma function. The logarithmic derivative

$$\Psi(z) = \frac{d}{dz}\log\Gamma(z) = \frac{\Gamma'(z)}{\Gamma(z)}$$

is known as the *digamma function*. Using the well-known relation

$$\Psi(x+1) - \Psi(y+1) = \sum_{n=1}^{\infty}\left(\frac{1}{n+y} - \frac{1}{n+x}\right), \tag{10.15}$$

we obtain

$$S_{d,p}(b) = \Psi\left(1 + \frac{P^+_{p,d}(1,b)}{1+b+\cdots+b^{p-1}}\right) - \Psi\left(1 + \frac{P_{p,d}(1,b)}{1+b+\cdots+b^{p-1}}\right).$$

In particular,

$$S_{1,1}(b) = \Psi(2) - \Psi(1+b) = 1 - \mathrm{C} - \Psi(1+b),$$

where C is *Euler's constant*. We thus have

Corollary 10.1. *For any* $0 < b < 1$,

$$\sum_{0<p/q<1} S_{p,q}(b) = \mathrm{C} + \Psi(1+b),$$

where the sum in the right-hand side extends over all irreducible fractions p/q *in the interval* $(0,1)$.

10.10 Hausdorff Dimension of $\Gamma_{a,b}$

We first assume that $0 < b \le a < 1$. Let us arrange all irreducible fractions in the interval $(0, 1)$ as follows:

$$\frac{1}{2}, \frac{1}{3}, \frac{2}{3}, \frac{1}{4}, \frac{3}{4}, \frac{1}{5}, \frac{2}{5}, \frac{3}{5}, \frac{4}{5}, \frac{1}{6}, \frac{5}{6}, \frac{1}{7}, \dots$$

where p/q lies ahead of r/s if $q < s$, or if $q = s$ and $p < r$. Denoting the nth irreducible fraction in the above arrangement by p_n/q_n it is easily seen that $p_n/q_n = 1/m$ when

$$n = \sum_{k=1}^{m-1} \varphi(k).$$

Thus the denominator q_n can be characterized as a unique integer satisfying

$$\sum_{k=1}^{q_n - 1} \varphi(k) \le n < \sum_{k=1}^{q_n} \varphi(k).$$

We now take $W = [0, 1]$ and

$$E_k = \operatorname{Int} \Delta\left(\frac{p_k}{q_k}\right)$$

for all $k \ge 1$. Then Theorem 10.7 implies that

$$\varepsilon_1 = \sum_{n=1}^{\infty} |E_n| = 1.$$

Moreover, using the inequality $\sqrt{n} < q_n$ we have

$$\begin{aligned}
\varepsilon_n = \sum_{k=n}^{\infty} |E_k| &\le \sum_{\substack{\gcd(p,q)=1 \\ q \ge q_n}} \left| \Delta\left(\frac{p}{q}\right) \right| \\
&\le \frac{1}{\lambda_0^2} \sum_{q \ge \sqrt{n}} \varphi(q) a^{q-2} \\
&\le \frac{1}{\lambda_0^2} \sum_{q \ge \sqrt{n}} q a^{q-2} \\
&< \frac{\sqrt{n}}{\lambda_0^4} a^{\sqrt{n}-1},
\end{aligned}$$

where λ_0 is the constant defined in Section 10.7. Therefore

$$\lim_{n \to \infty} \frac{\log n}{\log(n/\varepsilon_n)} = 0$$

and it follows from Theorem 8.8 that the Hausdorff dimension of the residual set $\Gamma_{a,b}$ is zero.

We next consider the case $0 < b < 1 \leq a$. Let $\{r_n/s_n\}$ be the upper approximation sequence associated with $\kappa_{a,b}$ and T_m, W_m be the same intervals as defined in Section 10.8. By the one-to-one correspondence

$$\frac{p}{q} \longleftrightarrow \frac{(q-p)r_m + pr_{m-1}}{(q-p)s_m + ps_{m-1}}$$

between all irreducible fractions in the interval $(0, 1)$ and that in $\operatorname{Int} W_m$, we can arrange all fractions in $\operatorname{Int} W_m$ using the ordering in the earlier discussion. So the nth fraction in the new ordering is

$$\frac{\widehat{p_n}}{\widehat{q_n}} = \frac{(q_n - p_n)r_m + p_n r_{m-1}}{(q_n - p_n)s_m + p_n s_{m-1}}$$

and we have

$$\begin{aligned}\widehat{q_n} &= (q_n - p_n)s_m + p_n s_{m-1} \\ &\geq s_{m-1} q_n \\ &> \sqrt{n}\, s_{m-1},\end{aligned}$$

because $s_m > s_{m-1}$ and $q_n > \sqrt{n}$. Put $W = T_m$ and

$$E_k = \operatorname{Int} \Delta\left(\frac{\widehat{p_k}}{\widehat{q_k}}\right)$$

for $k \geq 1$. Then we have

$$\varepsilon_n = \sum_{k=n}^{\infty} |E_k| \leq \sum_{\substack{p/q \in \operatorname{Int} W_m \\ q \geq \sqrt{n} s_{m-1}}} \left|\Delta\left(\frac{p}{q}\right)\right|.$$

Let ξ_m and λ_2 be the same constants as defined in Section 10.8. Since

$$\begin{aligned}\left|\Delta\left(\frac{p}{q}\right)\right| &\leq \frac{a^{q-p-1} b^{p-1}}{\lambda_2^2} \\ &\leq \frac{\xi_m^q}{ab\lambda_2^2},\end{aligned}$$

it follows that

$$\varepsilon_n \le \frac{1}{ab\lambda_2^2} \sum_{q \ge \sqrt{n} s_{m-1}} \varphi(q)\xi_m^q$$
$$< \frac{1}{ab\lambda_2^2} \sum_{q \ge \sqrt{n} s_{m-1}} q\xi_m^q$$
$$< \frac{s_{m-1}}{a\lambda_2^4 \xi_m} \sqrt{n}\,(\xi_m^{s_{m-1}})^{\sqrt{n}}.$$

Therefore

$$\lim_{n\to\infty} \frac{\log n}{\log(n/\varepsilon_n)} = 0,$$

and hence

$$\dim_{\mathrm{H}}(\Gamma_{a,b} \cap T_m) = 0$$

for all $m \ge 2$. By the countable stability property of the Hausdorff dimension we obtain

$$\dim_{\mathrm{H}} \Gamma_{a,b} = \sup_{m\ge 2} \dim_{\mathrm{H}}(\Gamma_{a,b} \cap T_m) = 0.$$

So rounding up the above arguments, we have the following

Theorem 10.10. *For any* $0 < b < 1$ *and* $a \ge b$ *we have* $\dim_{\mathrm{H}} \Gamma_{a,b} = 0$.

10.11 Rotation Number Functions

For arbitrarily fixed parameters a, b satisfying $0 < b < 1$ and $a \ge b$, it is natural to regard the rotation number $\rho(\psi_{a,b,c})$ as a function of c on the interval $(0, r_{a,b})$. In this sense we write

$$R_{a,b}(c) = \rho(\psi_{a,b,c}),$$

which we call the *rotation number function*. We have seen that $R_{a,b}(c)$ is monotonically decreasing on the set

$$\bigcup_{0<p/q<1} \Delta\left(\frac{p}{q}\right)$$

by Theorems 10.1 and 10.2. Moreover $R_{a,b}(c)$ is irrational if and only if $c \in \Gamma_{a,b}$ from Corollary 8.1 and Theorem 10.3. In such a case $\rho = R_{a,b}(c) \notin \mathbb{Q}$ satisfies

$$c = 1 - \frac{Q_{1/\rho}(b/a, a)}{a\,Q_\rho(a, b/a)}$$

by the basic formulae of the second kind (Theorems 8.6 and 8.7).

We now have the following

Theorem 10.11. *For any $0 < b < 1, a \geq b$ the rotation number function $R_{a,b}(c)$ is continuous and monotonically decreasing on the interval $(0, r_{a,b})$. Moreover $R_{a,b}(c)$ is singular; namely, the derivative of $R_{a,b}(c)$ exists and vanishes almost everywhere.*

Proof. For any $c \in \Gamma_{a,b}$, as with the proof of Theorem 10.3, let $a_n/b_n < c_n/d_n$ be two consecutive Farey fractions in $\mathfrak{F}_n \setminus \{0/1, 1/1\}$ such that $\Delta(a_n/b_n) = [\gamma_n, \delta_n]$ and $\Delta(c_n/d_n) = [\alpha_n, \beta_n]$ are the nearest Δ's on both sides of c. Since $|\Gamma_{a,b}| = 0$ by Theorems 10.7 and 10.8, the denominators b_n and d_n must tend to ∞ as $n \to \infty$. Since the proposition $\mathfrak{P}(a_n/b_n, c_n/d_n; \beta_n, \gamma_n)$ holds, it follows from the condition (2) in Section 10.1 that

$$U_k(\beta+) = U_k(c) = U_k(\gamma)$$

hold for $1 \leq k < b_n + d_n - 1$. So we have

$$\left\lceil \frac{c_n}{d_n}(b_n + d_n - 1) \right\rceil - 1 = \left[\rho(b_n + d_n - 1) \right]$$
$$= \left[\frac{a_n}{b_n}(b_n + d_n - 1) \right]$$

and hence

$$\lim_{n\to\infty} \frac{c_n}{d_n} = \rho = \lim_{n\to\infty} \frac{a_n}{b_n},$$

where $\rho = \rho(\psi_{a,b,c})$. This implies the continuity of the function $R_{a,b}(c)$ at each point of the residual set $\Gamma_{a,b}$ and the endpoints of Δ's. The singularity of $R_{a,b}(c)$ is straightforward from Theorems 10.7 and 10.8. □

Since it is easily verified that

$$\lim_{c\to 0+} R_{a,b}(c) = 1 \quad \text{and} \quad \lim_{c\to r_{a,b}-} R_{a,b}(c) = \begin{cases} 0 & (a \leq 1), \\ \kappa_{a,b} & (a > 1), \end{cases}$$

it is natural to extend $R_{a,b}(c)$ continuously to the endpoints of $[0, r_{a,b}]$.

The function $R_{a,b}(c)$ may evoke the well-known Cantor's function (Fig. 10.2), a standard example of singular functions. However, differing from Cantor's function, $R_{a,b}(c)$ does not satisfy any Hölder condition. To see this, for example, we take $x = 0$ and

$$y_q = \beta\left(\frac{q-1}{q}\right) = \frac{b^{q-1}}{1 + b + \cdots + b^{q-1}}.$$

We then have

$$\sup_{q\geq 2} \frac{\left|R_{a,b}(x) - R_{a,b}(y_q)\right|}{|x - y_q|^{\epsilon}} \geq \sup_{q\geq 2} \frac{1}{qb^{\epsilon(q-1)}} = \infty,$$

no matter how $\epsilon > 0$ is small.

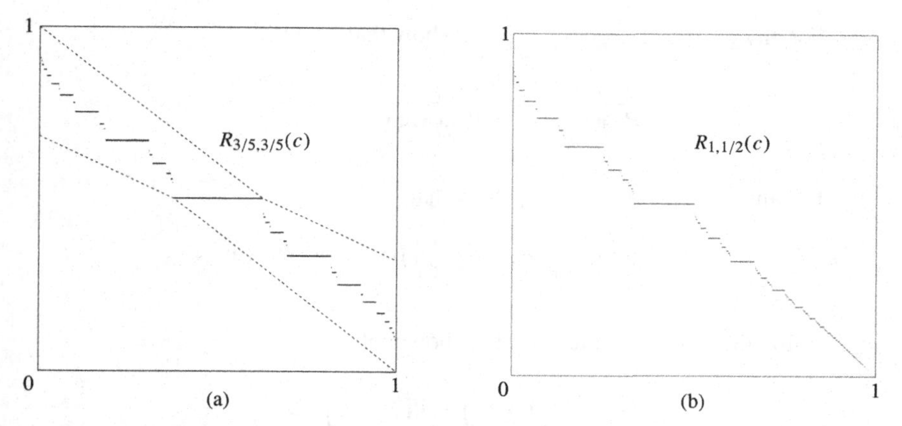

Fig. 10.1 (a) The graph of the rotation number function $R_{a,a}(c)$ with $a = 3/5$. The dotted lines indicate the upper and lower bounds given in (iii) of Exercise 6 in Chapter 9. (b) The graph of $R_{1,b}(c)$ with $b = 1/2$. Note that the left-hand derivative of $R_{1,b}(c)$ at $c = 1$ is -1 independently of b.

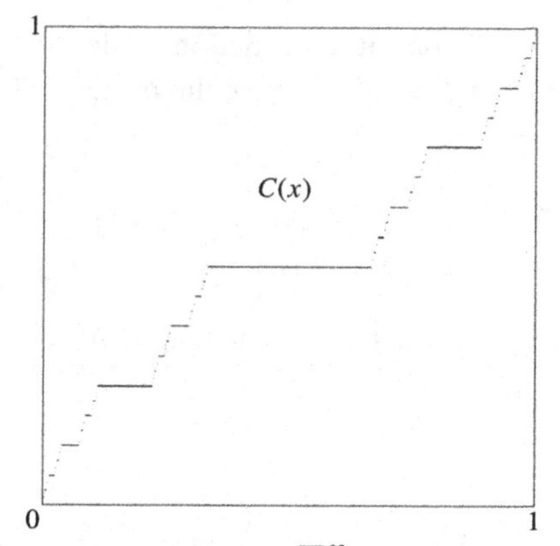

Fig. 10.2 The graph of Cantor's function $C(x) = \sum_{n=0}^{\infty} 2^{-n-1}\chi(\psi^n(x))$ where $\chi(x) = [(3x+1)/2]$ and $\psi(x) = \{3x - 3/2 - |3x-1|/2 + |3x-2|/2\}$. This is sometimes called the *Devil's staircase*.

Exercises in Chapter 10

1 For any irreducible fraction p/q in the interval $(0, 1)$ show that

$$P_{q,p}(1, z) + P^{+}_{q,q-p}(1, z) = z + z^2 + \cdots + z^q.$$

2 For any positive integers p, q and k show that

$$\frac{P_{kp,kq}(w, z)}{P_{p,q}(w, z)} = \frac{P^{+}_{kp,kq}(w, z)}{P^{+}_{p,q}(w, z)} = \frac{1 - w^{kq}z^{kp}}{1 - w^q z^p}.$$

3 For any positive integers p, q show that

$$w^q z^p\left(1 + \frac{1}{w}P_{p,q}\left(\frac{1}{w}, \frac{1}{z}\right)\right) = 1 + wzP^{+}_{p,q}(w, z).$$

4 Using the Lambert series (10.14) show that

$$\sum_{n=2}^{\infty} \frac{\varphi(n)}{n}\left(\Psi\left(1 + \frac{1}{n}\right) + \Psi\left(1 - \frac{1}{n}\right) + 2\mathrm{C}\right) = -1,$$

where $\varphi(n)$ is Euler's totient function, $\Psi(x)$ is the digamma function and C is Euler's constant.

5 Let E be the set of all continuous functions f defined on the interval $[0, 1]$ satisfying $f(0) = 0$ and $f(1) = 1$. Define the mapping $T : E \to E$ by

$$T(f)(x) = \begin{cases} f(3x)/2 & (0 \le x < 1/3), \\ 1/2 & (1/3 \le x < 2/3), \\ 1/2 + f(3x - 2)/2 & (2/3 \le x \le 1). \end{cases}$$

Show then that T is a contraction and its unique fixed point is Cantor's function $C(x)$.

Chapter 11

α- and β-Leaves

Apart from the parameter region Λ of the piecewise linear maps $\psi_{a,b,c}$, two kinds of rational functions of two variables are introduced and investigated in exact detail. The smooth surfaces defined by these functions are called the "leaves". To show fine hierarchical structures of the leaves we evaluate the following determinants:

$$\begin{vmatrix} P^+_{p,q}(w,z) & P_{q,p}(z,w) \\ P^+_{r,s}(w,z) & P_{s,r}(z,w) \end{vmatrix} \quad \text{and} \quad \begin{vmatrix} P_{p,q}(w,z) & P^+_{q,p}(z,w) \\ P_{r,s}(w,z) & P^+_{s,r}(z,w) \end{vmatrix}.$$

(See the formulae (11.1) and (11.6) respectively.)

11.1 Definition of Leaves

We regard the endpoints of $\Delta(p/q)$ as the functions of two real variables $w = a$ and $z = b/a$ by the formulae (10.3) and (10.4). So we define

$$\alpha_{p/q}(w,z) = 1 - \frac{P_{q,p}(z,w)}{wP^+_{p,q}(w,z)}$$

and

$$\beta_{p/q}(w,z) = 1 - \frac{P^+_{q,p}(z,w)}{wP_{p,q}(w,z)}$$

for any irreducible fraction p/q in the interval $(0,1)$. The surface defined by the rational function $\alpha_{p/q}(w,z)$ is said to be $\alpha_{p/q}$-leaf. Likewise $\beta_{p/q}$-leaf is meant the surface defined by the rational function $\beta_{p/q}(w,z)$. Obviously $\alpha_{p/q}(w,z) < 1$ and $\beta_{p/q}(w,z) < 1$ in the first quadrant†.

On the boundary of the first quadrant we have

$$\alpha_{p/q}(0,z) = \begin{cases} 1 & (p/q \le 1/2), \\ 0 & (p/q > 1/2), \end{cases} \qquad \alpha_{p/q}(w,0) = \frac{1 - w^{[q/p]-1}}{1 - w^{[q/p]}}$$

† The first quadrant is the open region $w, z > 0$ in the wz-plane.

and

$$\beta_{p/q}(0,z) = \begin{cases} 1 & (p/q < 1/2), \\ 0 & (p/q \geq 1/2), \end{cases} \qquad \beta_{p/q}(w,0) = \frac{1 - w^{\lceil q/p \rceil - 2}}{1 - w^{\lceil q/p \rceil - 1}}.$$

Thus $\alpha_{p/q}$- and $\beta_{p/q}$-leaves are smooth surfaces on the closed region $w, z \geq 0$.

The following lemmas, being straightforward consequences from the identities (10.5) and (10.6) respectively, state the hierarchical relationship of leaves in the first quadrant.

Lemma 11.1. *For any irreducible fraction p/q in the interval $(0, 1)$, $\alpha_{p/q}$-leaf is above or below $\beta_{p/q}$-leaf according to $w^q z^p < 1$ or $w^q z^p > 1$.*

Lemma 11.2. *For any irreducible fractions $p/q < r/s$ in $(0, 1)$, $\alpha_{r/s}$-leaf is below $\beta_{p/q}$-leaf.*

Two additional identities will be used to make the structure of leaves in the first quadrant clearer in Sections 11.2 and 11.3.

The following is known as Dini's theorem. We give the proof for completeness.

Theorem 11.1. *Let $\{f_n(x)\}$ be a sequence of real-valued continuous functions defined on a compact space X satisfying*

$$f_n(x) \leq f_{n+1}(x)$$

for all n and any $x \in X$. If $f_n(x)$ converges pointwisely to a continuous function $f(x)$ on X, then the convergence is uniform on X.

Proof. For any $\varepsilon > 0$ and $n \geq 1$ the set

$$E_n = \{x \in X : f(x) - f_n(x) < \varepsilon\}$$

is an open subset of X. Since each $x \in X$ belongs to E_n for all sufficiently large n, the sequence $\{E_n\}$ makes an open covering of X. Moreover E_n is monotonically increasing; that is, $E_n \subset E_{n+1}$ for all $n \geq 1$. Hence, by compactness, there exists a positive integer m satisfying $E_m = X$. This means that $f_n(x)$ converges uniformly to $f(x)$ on X. □

11.2 Limits of α-Leaves

Lemma 11.3. *For any irreducible fractions $p/q < r/s$ in $(0, 1)$, $\alpha_{p/q}$-leaf is above $\alpha_{r/s}$-leaf in the first quadrant.*

Proof. It suffices to show that $\alpha_{p/q}(w,z) > \alpha_{r/s}(w,z)$ under the condition $qr - ps = 1$. To this end we will show that the identity

$$P^+_{p,q}(w,z)\,P_{s,r}(z,w) - P_{q,p}(z,w)P^+_{r,s}(w,z) = w^{q-1}z^{p-1} \tag{11.1}$$

holds for any fractions p/q, r/s in the interval $[0,1]$ satisfying $qr - ps = 1$. Let $\mathfrak{P}(p/q, r/s)$ be the proposition that (11.1) holds identically. We first show the case $p/q = 0/1$; so, $r = 1$ and $s \geq 1$. Since $P_{s,1}(z,w) = w^s$,

$$P^+_{0,1}(w,z) = \frac{1}{z}, \quad P^+_{1,s}(w,z) = \frac{w^s - 1}{w - 1} \quad \text{and} \quad P_{1,0}(z,w) = \frac{w-1}{z},$$

the proposition $\mathfrak{P}(0/1, 1/s)$ certainly holds for all $s \geq 1$.

We next assume that $\mathfrak{P}(p/q, r/s)$ holds for some fractions p/q and r/s in the interval $[0,1]$ satisfying $qr - ps = 1$.

(a) We first deal with $\mathfrak{P}(p/q, (p+r)/(q+s))$. By the previous consideration we can assume that $p \geq 1$. Using the formulae (10.7) and (10.8) we have

$$\begin{aligned}
&P^+_{p,q}(w,z)\,P_{q+s,p+r}(z,w) - P_{q,p}(z,w)P^+_{p+r,q+s}(w,z)\\
&\quad = P^+_{p,q}(w,z)\,P_{s,r}(z,w) - P_{q,p}(z,w)P^+_{r,s}(w,z)\\
&\quad = w^{q-1}z^{p-1},
\end{aligned}$$

which means that the proposition $\mathfrak{P}(p/q, (p+r)/(q+s))$ is also true.

(b) Secondly we deal with $\mathfrak{P}((p+r)/(q+s), r/s)$. From the formulae (10.7) and (10.8) again, we have

$$\begin{aligned}
&P^+_{p+r,q+s}(w,z)\,P_{s,r}(z,w) - P_{q+s,p+r}(z,w)P^+_{r,s}(w,z)\\
&\quad = w^s z^r\big(P^+_{p,q}(w,z)\,P_{s,r}(z,w) - P_{q,p}(z,w)P^+_{r,s}(w,z)\big)\\
&\quad = w^{q+s-1}z^{p+r-1},
\end{aligned}$$

and hence the proposition $\mathfrak{P}((p+r)/(q+s), r/s)$ is also true.

Applying Lemma 5.4 we see that the proposition $\mathfrak{P}(p/q, r/s)$ is true for any fractions p/q, r/s in the interval $[0,1]$ satisfying $qr - ps = 1$. This completes the proof. □

Concerning convergence of α-leaves from above, we have

Lemma 11.4. *Let $\{p_n/q_n\}$ be a strictly monotonically increasing sequence of rational numbers converging to an irreducible fraction r/s in the interval $(0,1]$. Then the sequence of rational functions $\alpha_{p_n/q_n}(w,z)$ converges to $\alpha_{r/s}(w,z)$ when*

$w^s z^r \le 1$ and to

$$A^+_{r/s}(w,z) = 1 - \frac{1}{w} \cdot \frac{(w^s z^r - 1)P_{s',r'}(z,w) + P_{s,r}(z,w)}{(w^s z^r - 1)P^+_{r',s'}(w,z) + P^+_{r,s}(w,z)}$$

when $w^s z^r \ge 1$, where r'/s' is the left-predecessor of r/s if $s \ge 2$ and $r'/s' = 0/1$ if $r/s = 1/1$. The convergence is uniform on any compact set in the closed region $w, z \ge 0$. Moreover $A^+_{r/s}$-leaf is above $\alpha_{r/s}$-leaf in the region $w^s z^r > 1$.

Proof. By Lemma 11.3 we can assume that $q_n r - p_n s = 1$. Put $\ell_n = [q_n/s]$ for brevity. Since

$$\frac{r}{s}k - \frac{1}{s} < \frac{p_n}{q_n}k < \frac{r}{s}k$$

for $1 \le k < q_n$, we have $\lceil p_n k/q_n \rceil = \lceil rk/s \rceil$, and hence

$$P^+_{p_n,q_n}(w,z) = \sum_{k=1}^{q_n-1} w^{k-1} z^{\lceil rk/s \rceil - 1} + w^{q_n-1} z^{p_n-1}.$$

First, if $s \ge 2$, then $a_n = p_n - \ell_n r$ and $b_n = q_n - \ell_n s$ satisfy $0 \le a_n < r$ and $1 \le b_n < s$. Since $b_n r - a_n s = 1$, a_n/b_n is the left-predecessor of r/s by (i) of Lemma 5.7, and hence $a_n = r'$ and $b_n = s'$ by definition. Therefore

$$P^+_{p_n,q_n}(w,z) = \left(1 + w^s z^r + \cdots + (w^s z^r)^{\ell_n - 1}\right) \sum_{k=1}^{s} w^{k-1} z^{\lceil rk/s \rceil - 1}$$
$$+ (w^s z^r)^{\ell_n} \left(\sum_{k=1}^{s'-1} w^{k-1} z^{\lceil rk/s \rceil - 1} + w^{s'-1} z^{r'-1} \right).$$

Since $s' < s$, we have $\lceil rk/s \rceil = \lceil r'k/s' \rceil$ for $1 \le k < s'$. The last term $w^{s'-1} z^{r'-1}$ in the second line can be added into the previous sum as the term corresponding to $k = s'$ so that we get

$$P^+_{p_n,q_n}(w,z) = \left(1 + w^s z^r + \cdots + (w^s z^r)^{\ell_n - 1}\right) P^+_{r,s}(w,z)$$
$$+ (w^s z^r)^{\ell_n} P^+_{r',s'}(w,z). \qquad (11.2)$$

Secondly, if $s = 1$, then $r = 1$ and $q_n = p_n + 1$. We see that

$$P^+_{p_n,p_n+1}(w,z) = 1 + wz + \cdots + (wz)^{p_n-1} + w^{p_n} z^{p_n-1}$$

can be expressed in the form (11.2) with $\ell_n = p_n$, $r' = 0$ and $s' = 1$, because $P^+_{1,1}(w,z) = 1$ and $P^+_{0,1}(w,z) = 1/z$.

On the other hand, since

$$\frac{s}{r}k < \frac{q_n}{p_n}k < \frac{s}{r}k + \frac{1}{r}$$

for $1 \le k < p_n$, we have $[sk/r] = [q_n k/p_n]$. First, if $r \ge 2$, then $s \ge 3$, $r' \ge 1$ and

$$\begin{aligned} P_{q_n,p_n}(z,w) &= \sum_{k=1}^{p_n-1} w^{[sk/r]} z^{k-1} + w^{q_n} z^{p_n-1} \\ &= \left(1 + w^s z^r + \cdots + (w^s z^r)^{\ell_n - 1}\right) P_{s,r}(z,w) \\ &\quad + (w^s z^r)^{\ell_n} \left(\sum_{k=1}^{r'-1} w^{[sk/r]} z^{k-1} + w^{s'} z^{r'-1} \right). \end{aligned}$$

Since $r' < r$, $[sk/r] = [s'k/r']$ for $1 \le k < r'$ and the last term $w^{s'} z^{r'-1}$ in the third line can be added into the previous sum as the term corresponding to $k = r'$. We thus obtain

$$\begin{aligned} P_{q_n,p_n}(z,w) = \left(1 + w^s z^r + \cdots + (w^s z^r)^{\ell_n - 1}\right) P_{s,r}(z,w) \\ + (w^s z^r)^{\ell_n} P_{s',r'}(z,w). \qquad (11.3) \end{aligned}$$

Secondly, if $r = 1$, then $r' = 0$, $s' = 1$ and $q_n = p_n s + 1$. We see that

$$P_{p_n s+1,p_n}(z,w) = w^s (1 + w^s z + \cdots + (w^s z)^{p_n - 2}) + w^{p_n s+1} z^{p_n - 1}$$

can be expressed in the form (11.3) with $p_n = \ell_n$, because $P_{s,1}(z,w) = w^s$ and $P_{1,0}(z,w) = (w-1)/z$.

We now distinguish three cases as follows.

(a) $w^s z^r < 1$.

It follows from (11.2) and (11.3) that

$$\lim_{n\to\infty} P^+_{p_n,q_n}(w,z) = \frac{P^+_{r,s}(w,z)}{1 - w^s z^r} \quad \text{and} \quad \lim_{n\to\infty} P_{q_n,p_n}(z,w) = \frac{P_{s,r}(z,w)}{1 - w^s z^r}$$

respectively. Thus $\alpha_{p_n/q_n}(w,z)$ converges to $\alpha_{r/s}(w,z)$ as $n \to \infty$ when $w, z > 0$, while the convergence on the boundary is obvious.

(b) $w^s z^r = 1$.

In this case we have

$$P^+_{p_n,q_n}(w,z) = P^+_{r,s}(w,z)\,\ell_n + O(1) \quad \text{and} \quad P_{q_n,p_n}(z,w) = P_{s,r}(z,w)\,\ell_n + O(1)$$

as $n \to \infty$. Therefore $\alpha_{p_n/q_n}(w,z)$ converges to $\alpha_{r/s}(w,z)$ as $n \to \infty$.

(c) $w^s z^r > 1$.

It can be seen from (11.2) and (11.3) that

$$P^+_{p_n,q_n}(w,z) = \left(P^+_{r',s'}(w,z) + \frac{P^+_{r,s}(w,z)}{w^s z^r - 1}\right)(w^s z^r)^{\ell_n} + O(1),$$

$$P_{q_n,p_n}(z,w) = \left(P_{s',r'}(z,w) + \frac{P_{s,r}(z,w)}{w^s z^r - 1}\right)(w^s z^r)^{\ell_n} + O(1)$$

as $n \to \infty$. Hence $\alpha_{p_n/q_n}(w,z)$ converges to $A^+_{r/s}(w,z)$.

Since $\alpha_{r/s}(w,z) = A^+_{r/s}(w,z)$ on the curve $w^s z^r = 1$, $\alpha_{p_n/q_n}(w,z)$ converges pointwisely to a continuous function in the closed region $w, z \geq 0$. Therefore the convergence is uniform on any compact set by Theorem 11.1.

Finally, since r'/s' is the left-predecessor of r/s, it follows from Lemma 11.3 that $\alpha_{r'/s'}$-leaf is above $\alpha_{r/s}$-leaf. Hence $A^+_{r/s}$-leaf is also above $\alpha_{r/s}$-leaf when $w^s z^r > 1$. □

Concerning convergence of α-leaves from below, we have

Lemma 11.5. *Let $\{r_n/s_n\}$ be a strictly monotonically decreasing sequence of rational numbers converging to an irreducible fraction p/q in the interval $[0, 1)$. Then the sequence of rational functions $\alpha_{r_n/s_n}(w,z)$ converges to $\beta_{p/q}(w,z)$ when $w^q z^p \leq 1$ and to*

$$A^-_{p/q}(w,z) = 1 - \frac{1}{w} \cdot \frac{(w^q z^p - 1)P_{q',p'}(z,w) + P^+_{q,p}(z,w)}{(w^q z^p - 1)P^+_{p',q'}(w,z) + P_{p,q}(w,z)}$$

when $w^q z^p \geq 1$, where p'/q' is the right-predecessor of p/q if $q \geq 2$ and $p'/q' = 1/1$ if $p/q = 0/1$. The convergence is uniform on any compact set in the closed region $w, z \geq 0$. Moreover $A^-_{p/q}$-leaf is below $\alpha_{p/q}$-leaf in the region $w^q z^p > 1$.

Proof. We can assume that $qr_n - ps_n = 1$. Put $\ell_n = [s_n/q]$ for brevity. We have

$$\frac{p}{q}k < \frac{r_n}{s_n}k < \frac{p}{q}k + \frac{1}{q}$$

for $1 \leq k < s_n$. First, if $q \geq 2$, then $p \geq 1$ and

$$\left\lceil \frac{r_n}{s_n}k \right\rceil = \begin{cases} \left\lceil \frac{p}{q}k \right\rceil & (k \not\equiv 0 \pmod{q}), \\ \left\lceil \frac{p}{q}k \right\rceil + 1 & (k \equiv 0 \pmod{q}), \end{cases}$$

for $1 \leq k \leq s_n$. Then it is easily seen that $c_n = r_n - \ell_n p$ and $d_n = s_n - \ell_n q$ satisfy $1 \leq c_n \leq p$ and $1 \leq d_n < q$. Since $c_n q - d_n p = 1$, c_n/d_n is the right-predecessor of

p/q by (ii) of Lemma 5.7, and hence $c_n = p'$ and $d_n = q'$ by definition. Therefore

$$P^+_{r_n,s_n}(w,z) = \left(1 + w^q z^p + \cdots + (w^q z^p)^{\ell_n - 1}\right)\left(\sum_{k=1}^{q-1} w^{k-1} z^{\lceil pk/q \rceil - 1} + w^{q-1} z^p\right)$$
$$+ (w^q z^p)^{\ell_n} \sum_{k=1}^{q'} w^{k-1} z^{\lceil pk/q \rceil - 1}.$$

Replacing $\lceil pk/q \rceil - 1$ by $[pk/q]$ the last term $w^{q-1}z^p$ in the first line can be added into the previous sum as the term corresponding to $k = q$ so that the new sum becomes $P_{p,q}(w,z)$. Since $q' < q$, we get $\lceil pk/q \rceil = \lceil p'k/q' \rceil$ for $1 \le k \le q'$, and

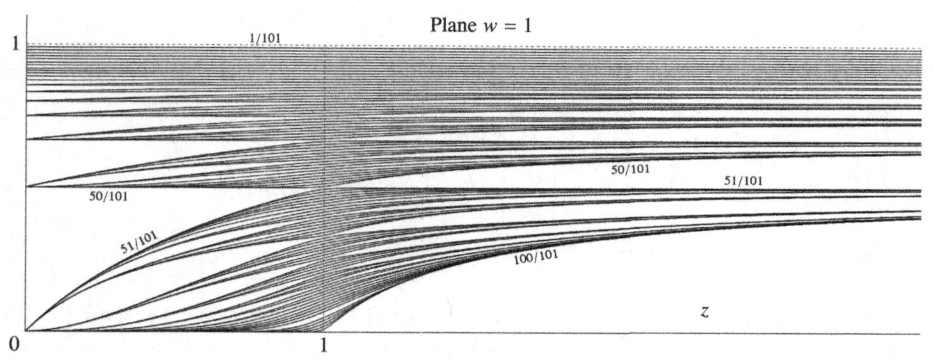

Fig. 11.1 The cross-section of α-leaves with the plane $w = 1$. One hundred curves ($1 \le p \le 100$, $q = 101$) are plotted.

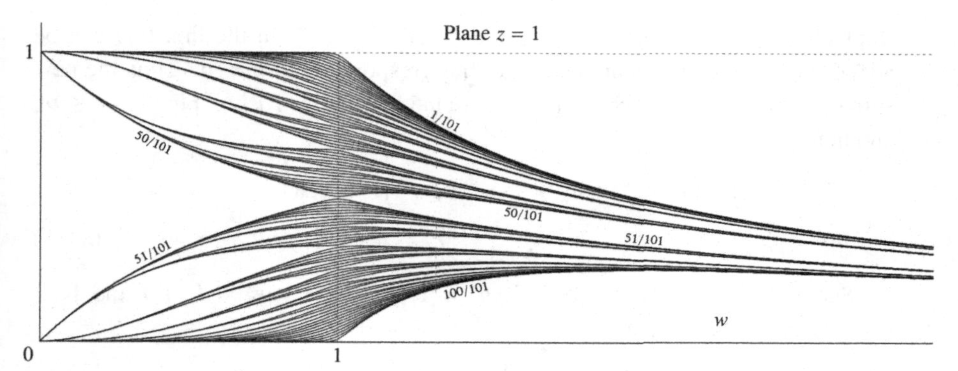

Fig. 11.2 The cross-section of α-leaves with the plane $z = 1$. One hundred curves ($1 \le p \le 100$, $q = 101$) are plotted.

thus

$$P^+_{r_n,s_n}(w,z) = \left(1 + w^q z^p + \cdots + (w^q z^p)^{\ell_n - 1}\right) P_{p,q}(w,z) \\ + (w^q z^p)^{\ell_n} P^+_{p',q'}(w,z). \qquad (11.4)$$

Secondly, if $q = 1$, then $p = 0$ and $r_n = 1$. We see that

$$P^+_{1,s_n}(w,z) = 1 + w + \cdots + w^{s_n - 1}$$

can be expressed in the form (11.4) with $\ell_n = s_n - 1$ and $p' = q' = 1$, because $P_{0,1}(w,z) = P^+_{1,1}(w,z) = 1$.

On the other hand,

$$\frac{q}{p}k - \frac{1}{p} < \frac{s_n}{r_n}k < \frac{q}{p}k$$

holds for $1 \le k < r_n$. First, if $p \ge 2$, then $q \ge 3$ and

$$\left[\frac{s_n}{r_n}k\right] = \begin{cases} \left[\frac{q}{p}k\right] & (k \not\equiv 0 \pmod p), \\ \left[\frac{q}{p}k\right] - 1 & (k \equiv 0 \pmod p), \end{cases}$$

for $1 \le k \le r_n$. Therefore,

$$P_{s_n,r_n}(z,w) = \left(1 + w^q z^p + \cdots + (w^q z^p)^{\ell_n - 1}\right)\left(\sum_{k=1}^{p-1} w^{[qk/p]} z^{k-1} + w^{q-1} z^{p-1}\right) \\ + (w^q z^p)^{\ell_n} \sum_{k=1}^{p'} w^{[qk/p]} z^{k-1}.$$

Replacing $[qk/p]$ by $\lceil qk/p \rceil - 1$ the last term $w^{q-1}z^{p-1}$ in the first line can be added into the previous sum as the term corresponding to $k = p$. Thus the new sum becomes $P^+_{q,p}(z,w)$. Since $p' < p$, we have $[qk/p] = [q'k/p']$ for $1 \le k \le p'$, and hence

$$P_{s_n,r_n}(z,w) = \left(1 + w^q z^p + \cdots + (w^q z^p)^{\ell_n - 1}\right) P^+_{q,p}(z,w) \\ + (w^q z^p)^{\ell_n} P_{q',p'}(z,w). \qquad (11.5)$$

Secondly, if $p = 1$, then $q \ge 2$, $p' = 1$, $q' = q - 1$, $r_n = \ell_n + 1$ and $s_n = (\ell_n + 1)q - 1$. We see that

$$P_{s_n,r_n}(z,w) = w^{q-1}\left(1 + w^q z + \cdots + (w^q z)^{\ell_n}\right)$$

can be expressed in the form (11.5), because $P^+_{q,1}(z,w) = P_{q-1,1}(z,w) = w^{q-1}$.

Thirdly, if $p = 0$, then $q = r_n = 1$ and

$$P_{s_n,1}(z, w) = w^{s_n}$$

can also be expressed in the form (11.5) with $\ell_n = s_n - 1$ and $p' = q' = 1$, because $P^+_{1,0}(z, w) = 0$ and $P_{1,1}(z, w) = w$.

We now distinguish three cases as follows.

(a) $w^q z^p < 1$.

It follows from (11.4) and (11.5) that

$$\lim_{n\to\infty} P^+_{r_n,s_n}(w, z) = \frac{P_{p,q}(w, z)}{1 - w^q z^p} \quad \text{and} \quad \lim_{n\to\infty} P_{s_n,r_n}(z, w) = \frac{P^+_{q,p}(z, w)}{1 - w^q z^p}$$

respectively. Thus $\alpha_{r_n/s_n}(w, z)$ converges to $\beta_{p/q}(w, z)$ as $n \to \infty$.

(b) $w^q z^p = 1$.

The same conclusion as (a) follows, because

$$P^+_{r_n,s_n}(w, z) = P_{p,q}(w, z)\,\ell_n + O(1) \quad \text{and} \quad P_{s_n,r_n}(z, w) = P^+_{q,p}(z, w)\,\ell_n + O(1)$$

as $n \to \infty$.

(c) $w^q z^p > 1$.

It follows from (11.4) and (11.5) that

$$P^+_{r_n,s_n}(w, z) = \left(P^+_{p',q'}(w, z) + \frac{P_{p,q}(w, z)}{w^q z^p - 1}\right)(w^q z^p)^{\ell_n} + O(1),$$

$$P_{s_n,r_n}(z, w) = \left(P_{q',p'}(z, w) + \frac{P^+_{q,p}(z, w)}{w^q z^p - 1}\right)(w^q z^p)^{\ell_n} + O(1)$$

as $n \to \infty$, and hence $\alpha_{r_n/s_n}(w, z)$ converges to $A^-_{p/q}(w, z)$.

Since $\beta_{p/q}(w, z) = A^-_{p/q}(w, z)$ on the curve $w^q z^p = 1$, $\alpha_{r_n/s_n}(w, z)$ converges pointwisely to a continuous function in the closed region $w, z \geq 0$. Therefore the convergence is uniform on any compact set by Theorem 11.1.

Finally, since p'/q' is the right-predecessor of p/q, it follows from Lemma 11.3 that $\alpha_{p'/q'}$-leaf is below $\alpha_{p/q}$-leaf. Hence $A^-_{p/q}$-leaf is also below $\alpha_{p/q}$-leaf when $w^q z^p > 1$. □

For example, it follows from Lemma 11.5 that the function

$$\begin{cases} \beta_{0/1}(w, z) = 1 & (0 \leq w < 1), \\ A^-_{0/1}(w, z) = \dfrac{1}{w} & (w \geq 1), \end{cases}$$

gives the upper envelope of α-leaves. Indeed, one can observe that the graph of $\alpha_{1/101}(w, 1)$ illustrated in Fig. 11.2 is very close to $1/w$ when $w \geq 1$. Similarly, from Lemma 11.4 the function

$$\begin{cases} \alpha_{1/1}(w, z) = 0 & (0 \leq wz < 1), \\ A^{+}_{1/1}(w, z) = \dfrac{wz - 1}{w(wz + z - 1)} & (wz \geq 1), \end{cases}$$

gives the lower envelope of α-leaves. The graph of $\alpha_{100/101}(w, 1)$ illustrated in Fig. 11.2 is close to $w^{-1} - w^{-2}$ when $w \geq 1$. In particular, all α-leaves lie between 0 and 1 in the first quadrant.

11.3 Limits of β-Leaves

We will show that β-leaves have the same regularity as α-leaves as follows.

Lemma 11.6. *For any irreducible fractions $p/q < r/s$ in $(0, 1)$, $\beta_{p/q}$-leaf is above $\beta_{r/s}$-leaf in the first quadrant.*

Proof. The proof is just like that of Lemma 11.3. Let $\mathfrak{P}(p/q, r/s)$ be the proposition that

$$P_{p,q}(w, z)\, P^{+}_{s,r}(z, w) - P^{+}_{q,p}(z, w) P_{r,s}(w, z) = w^{s-1} z^{r-1} \tag{11.6}$$

holds identically. Since $P_{0,1}(w, z) = P^{+}_{1,1}(z, w) = 1$ and $P^{+}_{1,0}(z, w) = 0$, the proposition $\mathfrak{P}(0/1, 1/1)$ certainly holds.

We next assume that $\mathfrak{P}(p/q, r/s)$ holds for some fractions p/q and r/s in the interval $[0, 1]$ satisfying $qr - ps = 1$.

(a) We first deal with $\mathfrak{P}(p/q, (p + r)/(q + s))$. Using the formulae (10.9) and (10.10) we have

$$\begin{aligned} &P_{p,q}(w, z)\, P^{+}_{q+s,p+r}(z, w) - P^{+}_{q,p}(z, w) P_{p+r,q+s}(w, z) \\ &\quad = w^{q} z^{p} \left(P_{p,q}(w, z)\, P^{+}_{s,r}(z, w) - P^{+}_{q,p}(z, w) P_{r,s}(w, z)\right) \\ &\quad = w^{q+s-1} z^{p+r-1}, \end{aligned}$$

which means that the proposition $\mathfrak{P}(p/q, (p + r)/(q + s))$ is true.

(b) Similarly we deal with $\mathfrak{P}((p + r)/(q + s), r/s)$. From the formulae (10.9) and (10.10) we have

$$\begin{aligned}P_{p+r,q+s}(w,z)\,P^{+}_{s,r}(z,w) - P^{+}_{q+s,p+r}(z,w)P_{r,s}(w,z)\\ = P_{p,q}(w,z)\,P^{+}_{s,r}(z,w) - P^{+}_{q,p}(z,w)P_{r,s}(w,z)\\ = w^{s-1}z^{r-1}.\end{aligned}$$

Hence the proposition $\mathfrak{P}((p+r)/(q+s), r/s)$ is also true.

This completes the proof. □

Just like Lemma 11.4 the following lemma describes convergence of β-leaves from above.

Lemma 11.7. *Let $\{p_n/q_n\}$ be a strictly monotonically increasing sequence of rational numbers converging to an irreducible fraction r/s in the interval $(0,1]$. Then the sequence of rational functions $\beta_{p_n/q_n}(w,z)$ converges to $\alpha_{r/s}(w,z)$ when $w^s z^r \le 1$ and to*

$$B^{+}_{r/s}(w,z) = 1 - \frac{1}{w}\cdot\frac{(w^s z^r - 1)P^{+}_{s',r'}(z,w) + P_{s,r}(z,w)}{(w^s z^r - 1)P_{r',s'}(w,z) + P^{+}_{r,s}(w,z)}$$

when $w^s z^r \ge 1$, where r'/s' is the left-predecessor of r/s if $s \ge 2$ and $r'/s' = 0/1$ if $r/s = 1/1$. The convergence is uniform on any compact set in the closed region $w, z \ge 0$. Moreover $B^{+}_{r/s}$-leaf is above $\beta_{r/s}$-leaf in the region $w^s z^r > 1$.

Proof. We can assume that $q_n r - p_n s = 1$. Put $\ell_n = [q_n/s]$ for brevity. First, if $s \ge 2$, then we have

$$\left[\frac{p_n}{q_n}k\right] = \begin{cases} \left[\dfrac{r}{s}k\right] & (k \not\equiv 0 \pmod{s}),\\[2ex] \left[\dfrac{r}{s}k\right] - 1 & (k \equiv 0 \pmod{s}),\end{cases}$$

for $1 \le k \le q_n$, and it can be seen that $a_n = p_n - \ell_n r$ and $b_n = q_n - \ell_n s$ satisfy $b_n r - a_n s = 1$ with $0 \le a_n < r$ and $1 \le b_n < s$. Hence a_n/b_n is the left-predecessor of r/s, and so, $a_n = r'$ and $b_n = s'$. Therefore

$$\begin{aligned}P_{p_n,q_n}(w,z) = \left(1 + w^s z^r + \cdots + (w^s z^r)^{\ell_n - 1}\right)\left(\sum_{k=1}^{s-1} w^{k-1}z^{[rk/s]} + w^{s-1}z^{r-1}\right)\\ + (w^s z^r)^{\ell_n}\sum_{k=1}^{s'} w^{k-1}z^{[rk/s]}.\end{aligned}$$

Replacing $[rk/s]$ by $\lceil rk/s\rceil - 1$ the last term $w^{s-1}z^{r-1}$ in the first line can be added into the previous sum as the term corresponding to $k = s$ so that the new sum becomes $P^+_{r,s}(w,z)$. Since $[rk/s] = [r'k/s']$ for $1 \le k \le s'$, we have

$$P_{p_n,q_n}(w,z) = \left(1 + w^s z^r + \cdots + (w^s z^r)^{\ell_n - 1}\right) P^+_{r,s}(w,z) + (w^s z^r)^{\ell_n} P_{r',s'}(w,z). \qquad (11.7)$$

Secondly, if $s = 1$, then $r = 1$ and $q_n = p_n + 1$. We see that

$$P_{p_n,p_n+1}(w,z) = 1 + wz + \cdots + (wz)^{p_n}$$

can be expressed in the form (11.7) with $\ell_n = p_n$, $r' = 0$ and $s' = 1$, because $P_{0,1}(w,z) = P^+_{1,1}(w,z) = 1$.

On the other hand, if $r \ge 2$, then $s \ge 3$, $r' \ge 1$ and

$$\left\lceil \frac{q_n}{p_n} k \right\rceil = \begin{cases} \left\lceil \frac{s}{r} k \right\rceil & (k \not\equiv 0 \pmod r), \\ \left\lceil \frac{s}{r} k \right\rceil + 1 & (k \equiv 0 \pmod r), \end{cases}$$

holds for $1 \le k \le p_n$. Therefore

$$P^+_{q_n,p_n}(z,w) = \left(1 + w^s z^r + \cdots + (w^s z^r)^{\ell_n - 1}\right)\left(\sum_{k=1}^{r-1} w^{\lceil sk/r \rceil - 1} z^{k-1} + w^s z^{r-1}\right) + (w^s z^r)^{\ell_n} \sum_{k=1}^{r'} w^{\lceil sk/r \rceil - 1} z^{k-1}.$$

Replacing $\lceil sk/r \rceil - 1$ by $[sk/r]$ the last term $w^s z^{r-1}$ can be added into the previous sum as the term corresponding to $k = r$; so, the new sum becomes $P_{s,r}(z,w)$. Since $\lceil sk/r \rceil = \lceil s'k/r' \rceil$ for $1 \le k \le r'$, we have

$$P^+_{q_n,p_n}(z,w) = \left(1 + w^s z^r + \cdots + (w^s z^r)^{\ell_n - 1}\right) P_{s,r}(z,w) + (w^s z^r)^{\ell_n} P^+_{s',r'}(z,w). \qquad (11.8)$$

Secondly, if $r = 1$, then $r' = 0$, $s' = 1$ and $q_n = p_n s + 1$. We see that

$$P^+_{p_n s+1,p_n}(z,w) = w^s (1 + w^s z + \cdots + (w^s z)^{p_n - 1})$$

can be expressed in the form (11.8) with $p_n = \ell_n$, because $P_{s,1}(z,w) = w^s$ and $P^+_{1,0}(z,w) = 0$.

We distinguish three cases as follows.

(a) $w^s z^r < 1$.

It follows from (11.7) and (11.8) that

$$\lim_{n\to\infty} P_{p_n,q_n}(w,z) = \frac{P^+_{r,s}(w,z)}{1 - w^s z^r} \quad \text{and} \quad \lim_{n\to\infty} P^+_{q_n,p_n}(z,w) = \frac{P_{s,r}(z,w)}{1 - w^s z^r}$$

respectively. Hence $\beta_{p_n/q_n}(w,z)$ converges to $\alpha_{r/s}(w,z)$ as $n \to \infty$.

(b) $w^s z^r = 1$.

The same conclusion as (a) follows, because

$$P_{p_n,q_n}(w,z) = P^+_{r,s}(w,z)\,\ell_n + O(1) \quad \text{and} \quad P^+_{q_n,p_n}(z,w) = P_{s,r}(z,w)\,\ell_n + O(1)$$

as $n \to \infty$.

(c) $w^s z^r > 1$.

From (11.7) and (11.8) we have

$$P_{p_n,q_n}(w,z) = \left(P_{r',s'}(w,z) + \frac{P^+_{r,s}(w,z)}{w^s z^r - 1}\right)(w^s z^r)^{\ell_n} + O(1),$$

$$P^+_{q_n,p_n}(z,w) = \left(P^+_{s',r'}(z,w) + \frac{P_{s,r}(z,w)}{w^s z^r - 1}\right)(w^s z^r)^{\ell_n} + O(1)$$

as $n \to \infty$, and hence $\beta_{p_n/q_n}(w,z)$ converges to $B^+_{r/s}(w,z)$.

Since $\alpha_{r/s}(w,z) = B^+_{r/s}(w,z)$ on the curve $w^s z^r = 1$, $\beta_{p_n/q_n}(w,z)$ converges pointwisely to a continuous function in the closed region $w, z \geq 0$. Therefore the convergence is uniform on any compact set by Theorem 11.1.

Finally, since r'/s' is the left-predecessor of r/s, it follows from Lemma 11.6 that $\beta_{r'/s'}$-leaf is above $\beta_{r/s}$-leaf. Hence $B^+_{r/s}$-leaf is also above $\beta_{r/s}$-leaf when $w^s z^r > 1$. □

Concerning convergence of β-leaves from below, we have

Lemma 11.8. *Let $\{r_n/s_n\}$ be a strictly monotonically decreasing sequence of rational numbers converging to an irreducible fraction p/q in the interval $[0, 1)$. Then the sequence of rational functions $\beta_{r_n/s_n}(w,z)$ converges to $\beta_{p/q}(w,z)$ when $w^q z^p \leq 1$ and to*

$$B^-_{p/q}(w,z) = 1 - \frac{1}{w} \cdot \frac{(w^q z^p - 1)P^+_{q',p'}(z,w) + P^+_{q,p}(z,w)}{(w^q z^p - 1)P_{p',q'}(w,z) + P_{p,q}(w,z)}$$

when $w^q z^p \geq 1$, where p'/q' is the right-predecessor of p/q if $q \geq 2$ and $p'/q' = 1/1$ if $p/q = 0/1$. The convergence is uniform on any compact set in the closed region $w, z \geq 0$. Moreover $B^-_{p/q}$-leaf is below $\beta_{p/q}$-leaf in the region $w^q z^p > 1$.

Proof. We can assume that $qr_n - ps_n = 1$. Put $\ell_n = [s_n/q]$ for brevity. Since $[pk/q] = [r_nk/s_n]$ for $1 \leq k < s_n$, we have

$$P_{r_n,s_n}(w,z) = \sum_{k=1}^{s_n-1} w^{k-1}z^{[pk/q]} + w^{s_n-1}z^{r_n}.$$

First, if $q \geq 2$, then $c_n = r_n - \ell_n p$ and $d_n = s_n - \ell_n q$ satisfy $1 \leq c_n \leq p$ and $1 \leq d_n < q$. Since $c_nq - d_np = 1$, c_n/d_n is the right-predecessor of p/q, and hence

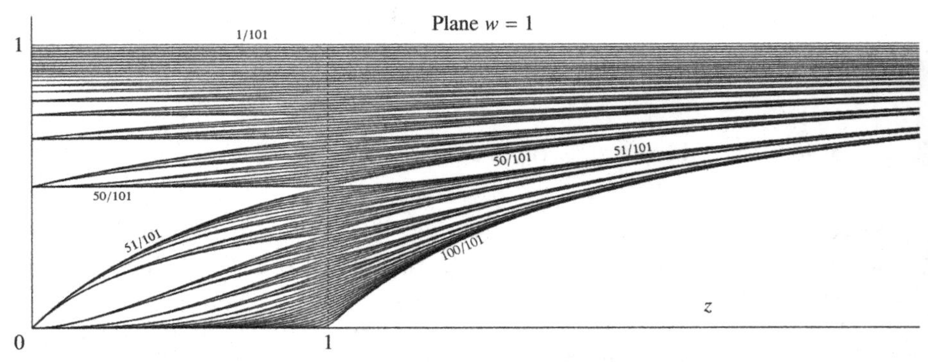

Fig. 11.3 The cross-section of β-leaves with the plane $w = 1$. One hundred curves ($1 \leq p \leq 100$, $q = 101$) are plotted.

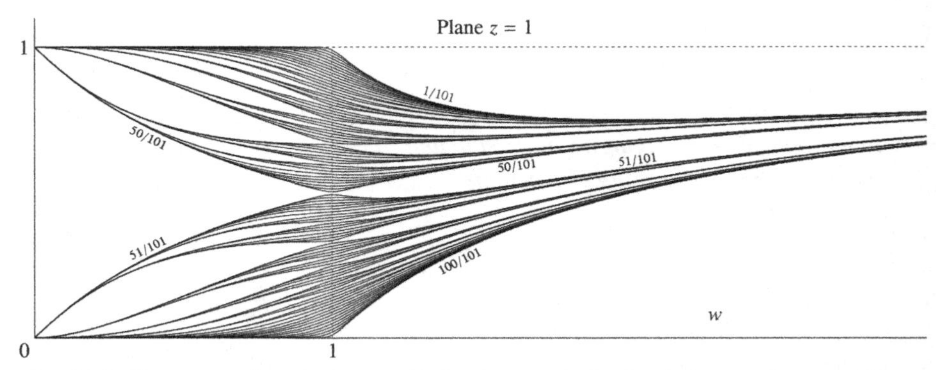

Fig. 11.4 The cross-section of β-leaves with the plane $z = 1$. One hundred curves ($1 \leq p \leq 100$, $q = 101$) are plotted.

$c_n = p'$ and $d_n = q'$. Therefore

$$P_{r_n,s_n}(w,z) = \left(1 + w^q z^p + \cdots + (w^q z^p)^{\ell_n - 1}\right) \sum_{k=1}^{q} w^{k-1} z^{[pk/q]}$$
$$+ (w^q z^p)^{\ell_n} \left(\sum_{k=1}^{q'-1} w^{k-1} z^{[pk/q]} + w^{q'-1} z^{p'} \right).$$

Since $q' < q$, we have $[pk/q] = [p'k/q']$ for $1 \le k < q'$; so, the last term $w^{q'-1}z^{p'}$ in the second line can be added into the previous sum as the term corresponding to $k = q'$. Hence

$$P_{r_n,s_n}(w,z) = \left(1 + w^q z^p + \cdots + (w^q z^p)^{\ell_n - 1}\right) P_{p,q}(w,z)$$
$$+ (w^q z^p)^{\ell_n} P_{p',q'}(w,z). \qquad (11.9)$$

Secondly, if $q = 1$, then $p = 0$ and $r_n = 1$. We see that

$$P_{1,s_n}(w,z) = 1 + w + \cdots + w^{s_n - 2} + w^{s_n - 1} z$$

can be expressed in the form (11.9) with $\ell_n = s_n - 1$, $p' = q' = 1$, because $P_{0,1}(w,z) = 1$ and $P_{1,1}(w,z) = z$.

On the other hand, if $p \ge 1$, then $\lceil s_n k / r_n \rceil = \lceil qk/p \rceil$ for $1 \le k < r_n$, and hence

$$P^+_{s_n,r_n}(z,w) = \sum_{k=1}^{r_n - 1} w^{\lceil qk/p \rceil - 1} z^{k-1} + w^{s_n - 1} z^{r_n - 1}$$
$$= \left(1 + w^q z^p + \cdots + (w^q z^p)^{\ell_n - 1}\right) P^+_{q,p}(z,w)$$
$$+ (w^q z^p)^{\ell_n} \left(\sum_{k=1}^{p'-1} w^{\lceil qk/p \rceil - 1} z^{k-1} + w^{q'-1} z^{p'-1} \right).$$

First, if $p \ge 2$, then $p' < p$ and we can replace $\lceil qk/p \rceil$ by $\lceil q'k/p' \rceil$. Thus the last term $w^{q'-1}z^{p'-1}$ in the third line can be added into the previous sum as the term corresponding to $k = p'$ so that the new sum becomes $P^+_{q',p'}(z,w)$. Therefore

$$P^+_{s_n,r_n}(z,w) = \left(1 + w^q z^p + \cdots + (w^q z^p)^{\ell_n - 1}\right) P^+_{q,p}(z,w)$$
$$+ (w^q z^p)^{\ell_n} P^+_{q',p'}(z,w). \qquad (11.10)$$

Secondly, if $p = 1$, then $q \ge 2$, $p' = 1$, $q' = q - 1$, $r_n = \ell_n + 1$ and $s_n = (\ell_n + 1)q - 1$. We see that

$$P^+_{s_n,r_n}(z,w) = w^{q-1}\left(1 + w^q z + \cdots + (w^q z)^{\ell_n - 1}\right) + (w^q z)^{\ell_n} w^{q-2}$$

can be expressed in the form (11.10), because $P^+_{q,1}(z,w) = w^{q-1}$ and $P^+_{q-1,1}(z,w) = w^{q-2}$.

Thirdly, if $p = 0$, then $q = r_n = 1$ and

$$P^+_{s_n,1}(z,w) = w^{s_n-1}$$

can be expressed in the form (11.10) with $\ell_n = s_n - 1$ and $p' = q' = 1$, because $P^+_{1,0}(z,w) = 0$ and $P^+_{1,1}(z,w) = 1$.

We now distinguish three cases as follows.

(a) $w^q z^p < 1$.

It follows from (11.9) and (11.10) that

$$\lim_{n\to\infty} P_{r_n,s_n}(w,z) = \frac{P_{p,q}(w,z)}{1 - w^q z^p} \quad \text{and} \quad \lim_{n\to\infty} P^+_{s_n,r_n}(z,w) = \frac{P^+_{q,p}(z,w)}{1 - w^q z^p}$$

respectively. Thus $\beta_{r_n/s_n}(w,z)$ converges to $\beta_{p/q}(w,z)$ as $n \to \infty$.

(b) $w^q z^p = 1$.

The same conclusion as (a) follows, because

$$P_{r_n,s_n}(w,z) = P_{p,q}(w,z)\ell_n + O(1) \quad \text{and} \quad P^+_{s_n,r_n}(z,w) = P^+_{q,p}(z,w)\ell_n + O(1)$$

as $n \to \infty$.

(c) $w^q z^p > 1$.

From (11.9) and (11.10) we get

$$P_{r_n,s_n}(w,z) = \left(P_{p',q'}(w,z) + \frac{P_{p,q}(w,z)}{w^q z^p - 1}\right)(w^q z^p)^{\ell_n} + O(1),$$

$$P^+_{s_n,r_n}(z,w) = \left(P^+_{q',p'}(z,w) + \frac{P^+_{q,p}(z,w)}{w^q z^p - 1}\right)(w^q z^p)^{\ell_n} + O(1)$$

as $n \to \infty$, and hence $\beta_{r_n/s_n}(w,z)$ converges to $B^-_{p/q}(w,z)$.

Since $\beta_{p/q}(w,z) = B^-_{p/q}(w,z)$ on the curve $w^q z^p = 1$, $\beta_{r_n/s_n}(w,z)$ converges pointwisely to a continuous function in the closed region $w, z \geq 0$. Therefore the convergence is uniform on any compact set by Theorem 11.1.

Finally, since p'/q' is the right-predecessor of p/q, it follows from Lemma 11.6 that $\beta_{p'/q'}$-leaf is below $\beta_{p/q}$-leaf. Hence $B^-_{p/q}$-leaf is also below $\beta_{p/q}$-leaf when $w^q z^p > 1$. □

It follows from Lemma 11.8 that the function

$$\begin{cases} \beta_{0/1}(w,z) = 1 & (0 \leq w < 1), \\ B^-_{0/1}(w,z) = \dfrac{w(w-1)z+1}{w(w-1)z+w} & (w \geq 1), \end{cases}$$

gives the upper envelope of β-leaves. Indeed, the graph of $\beta_{1/101}(w, 1)$ illustrated in Fig. 11.4 is very close to $1 - w^{-1} + w^{-2}$ when $w \geq 1$. Similarly, from Lemma 11.7 the function

$$\begin{cases} \alpha_{1/1}(w,z) = 0 & (0 \leq wz < 1), \\ B^+_{1/1}(w,z) = 1 - \dfrac{1}{wz} & (wz \geq 1), \end{cases}$$

gives the lower envelope of β-leaves. The graph of $\beta_{100/101}(1, z)$ illustrated in Fig. 11.3 is very close to $1 - z^{-1}$ when $z \geq 1$.

Exercises in Chapter 11

1 For any irreducible fraction p/q in the interval $(0,1)$ show that $\alpha_{p/q}(1,1) = \beta_{p/q}(1,1) = 1 - p/q$,

$$\lim_{z\to\infty} \alpha_{p/q}(1,z) = \frac{\lceil q/p\rceil - 1}{\lceil q/p\rceil} \quad \text{and} \quad \lim_{z\to\infty} \beta_{p/q}(1,z) = 1.$$

2 Show that each function $\alpha_{p/q}(1,z)$ is strictly monotonically increasing for $z \geq 0$ if $p \geq 2$. Some graphs of $\alpha_{p/q}(1,z)$ are illustrated in Fig. 11.1.

3 Show that each function $\beta_{p/q}(1,z)$ is strictly monotonically increasing for $z \geq 0$. Some graphs of $\beta_{p/q}(1,z)$ are illustrated in Fig. 11.3.

4 For any irreducible fraction p/q in the interval $(0,1)$ show that

$$\alpha_{p/q}(w,1) + \beta_{1-p/q}(w,1) = 1.$$

5 Show that the uniform norm of the difference between $\alpha_{1/q}(w,z)$ and the upper envelope function of α-leaves in the region $w, z \geq 0$ is less than or equal to $1/q$.

6 Show that the uniform norm of the difference between $\alpha_{p/(p+1)}(w,z)$ and the lower envelope function of α-leaves in the region $w, z \geq 0$ is less than or equal to $1/p$.

7 Show the identities (11.1) and (11.6) from (10.6) using the formulae

$$P^+_{p,q}(w,z) = P_{p,q}(w,z) + w^{q-1}z^{p-1}(1-z)$$

and

$$(1-w)P_{p,q}(w,z) + (1-z)P^+_{q,p}(z,w) = 1 - w^q z^p$$

stated in Exercises 1 and 2 in Chapter 8 respectively.

Chapter 12

Approximations to Hecke-Mahler Series

Using the hierarchical structure of the leaves investigated in the previous chapter we can give good rational approximations to the Hecke-Mahler series. Consequently we obtain several arithmetical properties of the values of the series at rational points. As another application of our approximations, we can show that the graph of the series $H_{1/\varphi}(w)$ oscillates infinitely often in the right-neighborhood of $w = -1$, where φ is the golden ratio.

12.1 Minimal Indices

Given non-zero formal power series

$$f(x,y) = \sum_{m,n\geq 0} c_{m,n} x^m y^n$$

of two variables x, y, let μ_0 be the smallest integer satisfying $c_{\mu_0,n} \neq 0$ for some $n \geq 0$, and ν_0 be the smallest integer satisfying $c_{m,\nu_0} \neq 0$ for some $m \geq 0$. In other words, μ_0 is the largest integer satisfying $c_{m,n} = 0$ for all $0 \leq m < \mu_0$, $n \geq 0$, and ν_0 is the largest integer satisfying $c_{m,n} = 0$ for all $0 \leq n < \nu_0$, $m \geq 0$. The pair of non-negative integers $\{\mu_0, \nu_0\}$ is called the *minimal indices* of $f(x,y)$.

We first need the following

Lemma 12.1. *Given an absolutely convergent series* $\sum_{m,n\geq 0} c_{m,n}$, *put*

$$f(x,y) = \sum_{m,n\geq 0} c_{m,n} x^m y^n.$$

Suppose that $f(x,y) \not\equiv 0$ *and*

$$\sup_{0<x,y\leq 1} \frac{|f(x,y)|}{x^M y^N} < \infty \tag{12.1}$$

for some non-negative integers M *and* N. *Then we have* $\mu_0 \geq M$ *and* $\nu_0 \geq N$ *where* $\{\mu_0, \nu_0\}$ *is the minimal indices of* $f(x,y)$.

Proof. Let ν^* be the smallest integer satisfying $c_{\mu_0,\nu^*} \neq 0$. For any $\kappa \geq \nu^* + 1$ we denote by ℓ_κ the straight line defined by

$$\kappa x + y = \kappa\mu_0 + \nu^*,$$

which has the negative slope $-\kappa$ and passes through the point (μ_0, ν^*). For any $(m,n) \neq (\mu_0, \nu^*)$ with $c_{m,n} \neq 0$, we have $m \geq \mu_0$ and $n \geq \nu_0$ by definition. If $m = \mu_0$, then $n > \nu^*$, and so $\kappa m + n \geq \kappa\mu_0 + \nu^* + 1$. On the other hand, if $m > \mu_0$, then

$$\begin{aligned} \kappa m + n &\geq \kappa(\mu_0 + 1) \\ &\geq \kappa\mu_0 + \nu^* + 1. \end{aligned}$$

Thus (μ_0, ν^*) is a unique lattice point with $c_{m,n} \neq 0$ lying on the line ℓ_κ, because all the other lattice points (m,n) with $c_{m,n} \neq 0$ are above the line ℓ_κ. Since

$$\begin{aligned} f(y^\kappa, y) &= \sum_{c_{m,n} \neq 0} c_{m,n} y^{\kappa m+n} \\ &= c_{\mu_0,\nu^*} y^{\kappa\mu_0+\nu^*} + \sum_{\substack{c_{m,n} \neq 0 \\ (m,n) \neq (\mu_0,\nu^*)}} c_{m,n} y^{\kappa m+n}, \end{aligned}$$

it follows that

$$\frac{|f(y^\kappa, y)|}{y^{\kappa M+N}} \geq \left(|c_{\mu_0,\nu^*}| - y \sum_{m,n \geq 0} |c_{m,n}| \right) y^{\kappa(\mu_0 - M) + \nu^* - N}$$

for any $0 < y < 1$. Suppose now that $\mu_0 < M$. Then we would have

$$\lim_{y \to 0+} \frac{|f(y^\kappa, y)|}{y^{\kappa M+N}} = \infty,$$

because

$$\kappa(\mu_0 - M) + \nu^* - N \leq -\kappa + \nu^* \leq -1.$$

This contradicts the assumption (12.1). Therefore we have $\mu_0 \geq M$. Interchanging x and y the same argument as above implies that $\nu_0 \geq N$. This completes the proof. □

12.2 Approximations to M_μ and H_μ

For any irrational number $\mu > 1$ let $\{p_n/q_n\}$ and $\{r_n/s_n\}$ be the lower and the upper approximation sequences associated with $1/\mu$ respectively. The Farey intervals $[p_n/q_n, r_n/s_n)$ make a monotonically descending sequence converging to $1/\mu$.

Let w, z be arbitrarily fixed positive numbers satisfying $w^\mu z < 1$. Then the function $Q_{1/\mu}(w,z)$ in (8.6) is well-defined, as well as $Q_\mu(z,w)$. Note that $w^{q_n}z^{p_n} < 1$ and $w^{s_n}z^{r_n} < 1$ for all sufficiently large n. Since $\{r_n/s_n\}$ is strictly monotonically decreasing and $\{p_n/q_n\}$ contains a strictly monotonically increasing subsequence, it follows from Lemma 8.2 with the ensuing note and Lemmas 11.3, 11.6 that

$$\alpha_{r_n/s_n}(w,z) < 1 - \frac{Q_\mu(z,w)}{wQ_{1/\mu}(w,z)} < \beta_{p_n/q_n}(w,z),$$

and hence

$$\frac{P^+_{q_n,p_n}(z,w)}{P_{p_n,q_n}(w,z)} < \frac{Q_\mu(z,w)}{Q_{1/\mu}(w,z)} < \frac{P_{s_n,r_n}(z,w)}{P^+_{r_n,s_n}(w,z)} \tag{12.2}$$

for all sufficiently large n.

We now distinguish three cases as follows.

(a) $0 < z < 1$.

Using the formulae stated in (8.5), Exercises 2 and 3 in Chapter 8, it follows from (12.2) that

$$\frac{wzP^+_{r_n,s_n}(w,z)}{(1-z)(1-w^{s_n}z^{r_n})} < M_\mu(w,z) < \frac{wzP_{p_n,q_n}(w,z)}{(1-z)(1-w^{q_n}z^{p_n})} \tag{12.3}$$

for $0 < w < z^{-1/\mu}$. In particular, putting $w = 1$ we obtain

$$\frac{zP^+_{r_n,s_n}(1,z)}{(1-z)(1-z^{r_n})} < H_\mu(z) < \frac{zP_{p_n,q_n}(1,z)}{(1-z)(1-z^{p_n})} \tag{12.4}$$

for $0 < z < 1$.

(b) $z = 1$.

Putting $z = 1$ in (12.2) and using the formula in Exercise 2 in Chapter 8, we have

$$\frac{wP^+_{q_n,p_n}(1,w)}{(1-w)(1-w^{q_n})} < H_{1/\mu}(w) < \frac{wP_{s_n,r_n}(1,w)}{(1-w)(1-w^{s_n})}$$

for $0 < w < 1$.

(c) $z > 1$.

Exactly as (a) we get

$$\frac{wzP_{p_n,q_n}(w,z)}{(z-1)(1-w^{q_n}z^{p_n})} < M_\mu(w,z) < \frac{wzP^+_{r_n,s_n}(w,z)}{(z-1)(1-w^{s_n}z^{r_n})}$$

for $0 < w < z^{-1/\mu}$.

To evaluate the errors of the above approximations we need the following

Lemma 12.2. *For any fractions p/q and r/s in the interval $[0, \infty)$ satisfying $qr - ps = 1$ we have the identity:*

$$(1 - w^s z^r)P_{p,q}(w, z) - (1 - w^q z^p)P^+_{r,s}(w, z) = (1 - z)w^{q+s-1}z^{p+r-1}. \tag{12.5}$$

Proof. Since either $p \geq q$ or $r \leq s$ holds, we can assume that $r/s \leq 1$ because of the formula stated in Exercise 2 in Chapter 8. Let $\mathfrak{P}(p/q, r/s)$ be the proposition that (12.5) holds identically. Since $P_{0,1}(w, z) = P^+_{1,1}(z, w) = 1$, the proposition $\mathfrak{P}(0/1, 1/1)$ certainly holds.

We next assume that $\mathfrak{P}(p/q, r/s)$ holds for some fractions p/q and r/s in the interval $[0, 1]$ satisfying $qr - ps = 1$.

(a) We first deal with $\mathfrak{P}(p/q, (p+r)/(q+s))$. Then using (10.8) and the formula stated in Exercise 1 in Chapter 8, we have

$$\begin{aligned}(1 - w^{q+s}z^{p+r})P_{p,q}(w, z) &- (1 - w^q z^p)P^+_{p+r,q+s}(w, z)\\ &= (1 - w^s z^r)P_{p,q}(w, z) - (1 - w^q z^p)P^+_{r,s}(w, z)\\ &\qquad - (1 - z)(1 - w^q z^p)w^{q+s-1}z^{p+r-1}\\ &= (1 - z)w^{2q+s-1}z^{2p+r-1},\end{aligned}$$

which means that the proposition $\mathfrak{P}(p/q, (p + r)/(q + s))$ is true.

(b) Similarly we deal with $\mathfrak{P}((p + r)/(q + s), r/s)$. From (10.9) and the formula stated in Exercise 1 in Chapter 8, we have

$$\begin{aligned}(1 - w^s z^r)P_{p+r,q+s}(w, z) &- (1 - w^{q+s}z^{p+r})P^+_{r,s}(w, z)\\ &= (1 - w^s z^r)P_{p,q}(w, z) - (1 - w^q z^p)P^+_{r,s}(w, z)\\ &\qquad - (1 - z)(1 - w^s z^r)w^{q+s-1}z^{p+r-1}\\ &= (1 - z)w^{q+2s-1}z^{p+2r-1},\end{aligned}$$

and hence the proposition $\mathfrak{P}((p + r)/(q + s), r/s)$ is also true.

This completes the proof. □

Theorem 12.1. *Let $\mu > 1$ be irrational and $\{p_n/q_n\}$, $\{r_n/s_n\}$ be the lower and the upper approximation sequences associated with $1/\mu$ respectively. Then the minimal indices $\{\mu_n, \nu_n\}$ of the power series*

$$f_n(x, y) = (1 - y)(1 - x^{q_n}y^{p_n})M_\mu(x, y) - xyP_{p_n,q_n}(x, y)$$

satisfies $\mu_n \geq q_n + s_n$ *and* $\nu_n \geq p_n + r_n$ *for all* $n \geq 1$.

Proof. For any $0 < x, y < 1$ it follows from (12.3) and Lemma 12.2 that

$$\begin{aligned}|f_n(x,y)| &< \frac{1-y}{1-x^{s_n}y^{r_n}} x^{q_n+s_n} y^{p_n+r_n} \\ &< x^{q_n+s_n} y^{p_n+r_n}.\end{aligned}$$

Since the series

$$\phi(x,y) = f_n\left(\frac{x}{2}, \frac{y}{2}\right)$$

converges absolutely for $|x|, |y| \leq 1$ and satisfies

$$|\phi(x,y)| < \frac{x^{q_n+s_n} y^{p_n+r_n}}{2^{p_n+q_n+r_n+s_n}} < x^{q_n+s_n} y^{p_n+r_n},$$

it follows from Lemma 12.1 that

$$\mu_n \geq q_n + s_n \quad \text{and} \quad \nu_n \geq p_n + r_n,$$

because the minimal indices of ϕ coincides with that of f_n. □

Similarly we have from (12.4) the following

Theorem 12.2. *Under the same assumptions as in the previous theorem,*

$$(1-x)(1-x^{p_n})H_\mu(x) - xP_{p_n,q_n}(1,x) = O(x^{p_n+r_n}) \tag{12.6}$$

as $x \to 0$ *for all* $n \geq 1$.

In general, for a given power series $f \in \mathbb{C}[[x]]$, in order to find two polynomials $A(x), B(x)$ in such a way that the coefficients of x^k of the series

$$A(x)f(x) - B(x)$$

vanish for all $0 \leq k < N$, one solves a system of N homogeneous linear equations with $M = \deg A + \deg B + 2$ variables. So, if $N < M$, then there exist always non-trivial solutions of the system. In this sense the formula (12.6) is marvelous, because

$$\begin{aligned}N - M &\geq p_n + r_n - (2p_n + 4) \\ &= r_{n-1} - 4\end{aligned}$$

tends to ∞ as $n \to \infty$.

12.3 Irrationality Exponents

Let γ be an irrational number. The *irrationality exponent* $\omega(\gamma)$ of γ is defined by

$$\omega(\gamma) = -\liminf_{q\to\infty} \min_{p\in\mathbb{Z}} \frac{\log|\gamma - p/q|}{\log q}.$$

In other words, $\omega(\gamma)$ is the least number ω such that, for any $\varepsilon > 0$ there exists a positive integer q_0 satisfying

$$\left|\gamma - \frac{p}{q}\right| > \frac{1}{q^{\omega+\varepsilon}}$$

for all $p \in \mathbb{Z}$ and all $q > q_0$. It is known that $\omega(\gamma) \in [2, \infty]$ and $\omega(\gamma) = 2$ for almost all irrational numbers γ. The irrational number γ with $\omega(\gamma) = \infty$ is said to be a Liouville number. In 1955 K. F. Roth showed that every algebraic irrational has the irrationality exponent 2. As an equivalent statement of Roth's theorem, γ is transcendental if $\omega(\gamma) > 2$. This gives a criterion for transcendency easy to use, while the set of irrational numbers satisfying $\omega(\gamma) > 2$ is a null set.

Roughly speaking, the irrationality exponent of γ measures its "nearness" to the rational numbers. For example, Liouville numbers are extremely close to the rational numbers, while algebraic irrational numbers are enormously far from the rational numbers.

Concerning arithmetical properties of the values of Hecke-Mahler series M_μ, it has already shown in Nishioka (1996) that $M_\mu(w, z)$, as well as $H_\mu(w)$, is transcendental for any irrational μ and for any non-zero algebraic numbers w, z with $|z| < 1$ and $|w^\mu z| < 1$. The following theorem gives lower estimates of irrationality exponents for such values in some specific cases, meaning that they are "rare" transcendental numbers.

Theorem 12.3. *Suppose that $\mu > 1$ is an irrational number with*

$$\vartheta = \limsup_{n\to\infty} \frac{s_n}{q_n} > 1,$$

where $\{q_n\}$ and $\{s_n\}$ are the denominators of the lower and the upper approximation sequences associated with $1/\mu$ respectively. Suppose further that k'/k and m'/m are irreducible fractions in the intervals $(0, 1]$ and $(0, 1)$ respectively satisfying

$$\frac{\mu \log k' + \log m'}{\mu \log k + \log m} < \frac{\vartheta - 1}{\vartheta + 1}. \tag{12.7}$$

Then $M_\mu(k'/k, m'/m)$ is a transcendental number having irrationality exponent *greater than or equal to*

$$(\vartheta + 1)\left(1 - \frac{\mu \log k' + \log m'}{\mu \log k + \log m}\right).$$

Proof. Put

$$\gamma = M_\mu\left(\frac{k'}{k}, \frac{m'}{m}\right)$$

for brevity. By substituting $w = k'/k$ and $z = m'/m$ into the formula (12.3) it follows from Lemma 12.2 that

$$0 < \left|\gamma - \frac{a_n}{b_n}\right| < 2\left(\frac{k'}{k}\right)^{q_n+s_n}\left(\frac{m'}{m}\right)^{p_n+r_n} \tag{12.8}$$

for all sufficiently large n, where

$$a_n = k'm' \cdot k^{q_n} m^{p_n} P_{p_n,q_n}\left(\frac{k'}{k}, \frac{m'}{m}\right),$$
$$b_n = k(m - m')(k^{q_n} m^{p_n} - k'^{q_n} m'^{p_n})$$

are positive integers.

For any $\varepsilon \in (0, 1/\mu)$ take a sufficiently large integer n_0 so that both the formula (12.8) and

$$p_n > \frac{1}{\varepsilon}\max\left(1, \frac{\log 2}{\log(m/m')}\right)$$

are valid for all $n \geq n_0$. Let $n_0 < n_1 < \cdots$ be a subsequence of integers satisfying

$$\frac{s_{n_j}}{q_{n_j}} > \vartheta - \varepsilon$$

for all $j \geq 1$. We write $p_{n_j} = p_j^*$, $q_{n_j} = q_j^*$, $r_{n_j} = r_j^*$, $s_{n_j} = s_j^*$ and $a_{n_j} = a_j^*$, $b_{n_j} = b_j^*$ for $j \geq 1$ shortly.

Since $p_n < q_n/\mu$ and $q_n > p_n > 1/\varepsilon$ for $n \geq n_0$, we obtain

$$1 \leq b_n < k^{q_n+1} m^{p_n+1} < \sigma^{q_n},$$

where $\sigma = (km^{1/\mu})^{1+\varepsilon}$. On the other hand, the right-hand side of (12.8) is less than

$$\left(\frac{k'}{k}\right)^{q_n+s_n}\left(\frac{m'}{m}\right)^{(1-\varepsilon)(p_n+r_n)}.$$

Since $r_n > s_n/\mu$ and $p_n > (1/\mu - \varepsilon)q_n$[†], we have

$$0 < \left|\gamma - \frac{a_j^*}{b_j^*}\right| < \tau^{-q_j^*}, \qquad (12.9)$$

where

$$\tau = \left(\frac{k}{k'}\right)^{1+\vartheta-\varepsilon}\left(\frac{m}{m'}\right)^{(1-\varepsilon)((1+\vartheta-\varepsilon)/\mu-\varepsilon)} > 1.$$

The formula (12.9) implies the irrationality of γ. To see this, suppose, on the contrary, that $\gamma = M/N$. Then it follows from (12.9) that

$$\frac{1}{N\sigma^{q_j^*}} < \frac{1}{Nb_j^*} \leq \left|\frac{M}{N} - \frac{a_j^*}{b_j^*}\right| < \tau^{-q_j^*},$$

which implies that $\sigma \geq \tau$, because q_j^* tends to ∞ as $j \to \infty$. However, under the condition (12.7) we have $\tau > \sigma^2$ for a sufficiently small ε, a contradiction.

More precisely, putting $\xi = \log\tau/\log\sigma$, we have from (12.9)

$$0 < \left|\gamma - \frac{a_j^*}{b_j^*}\right| < \frac{1}{b_j^{*\xi}},$$

and therefore

$$\min_{p\in\mathbb{Z}} \frac{\log\left|\gamma - p/b_j^*\right|}{\log b_j^*} < -\xi.$$

Since b_j^* tends to ∞ as $j \to \infty$, we get $\omega(\gamma) \geq \xi$, where

$$\xi = (\vartheta + 1)\left(1 - \frac{\mu\log k' + \log m'}{\mu\log k + \log m}\right) + O(\varepsilon)$$

as $\varepsilon \to 0+$. Since ε is arbitrary, this completes the proof. □

As a straightforward corollary, we have

Corollary 12.1. *Suppose that μ satisfies the same condition stated in Theorem 12.3. Then for any integers $k, m \geq 2$, $M_\mu(1/k, 1/m)$ and $H_\mu(1/m)$ are transcendental numbers having irrationality exponents greater than or equal to $\vartheta + 1$. Moreover, if $\vartheta = \infty$, then $M_\mu(k'/k, m'/m)$ and $H_\mu(m'/m)$ are Liouville numbers for all fractions k'/k and m'/m in the interval* $(0, 1)$.

We say that an irrational number γ has an irrationality measure[‡] ω when ω

[†] This follows from $|1/\mu - p_n/q_n| < 1/q_n < \varepsilon$.

[‡] This notion bears no relation to Measure Theory.

is an upper bound of the irrationality exponent $\omega(\gamma)$. Concerning irrationality measures for such values, we have

Theorem 12.4. *Suppose that $\mu > 1$ is an irrational number satisfying*

$$\vartheta = \limsup_{n\to\infty} \frac{s_n}{q_n} < \infty \quad \textit{and} \quad \vartheta^* = \limsup_{n\to\infty} \frac{s_{n+1}}{s_n} < \infty,$$

where $\{q_n\}$ and $\{s_n\}$ are the denominators of the lower and the upper approximation sequences associated with $1/\mu$ respectively. Suppose further that k'/k and m'/m are irreducible fractions in the intervals $(0, 1]$ and $(0, 1)$ respectively satisfying

$$\frac{\mu \log k' + \log m'}{\mu \log k + \log m} < \frac{1}{\vartheta + 1}. \tag{12.10}$$

Then $M_\mu(k'/k, m'/m)$ has an irrationality measure

$$1 + \vartheta\vartheta^* \left(1 - (\vartheta + 1)\frac{\mu \log k' + \log m'}{\mu \log k + \log m}\right)^{-1}.$$

Proof. Just like the proof of Theorem 12.3, putting

$$\gamma = M_\mu\left(\frac{k'}{k}, \frac{m'}{m}\right)$$

and substituting $w = k'/k$ and $z = m'/m$ into the formula (12.3), we obtain the formula (12.8) as well as

$$0 < \left|\gamma - \frac{c_n}{d_n}\right| < 2\left(\frac{k'}{k}\right)^{q_n+s_n}\left(\frac{m'}{m}\right)^{p_n+r_n}, \tag{12.11}$$

for all sufficiently large n, where

$$c_n = k'm' \cdot k^{s_n} m^{r_n} P^+_{r_n, s_n}\left(\frac{k'}{k}, \frac{m'}{m}\right),$$

$$d_n = k(m - m')(k^{s_n} m^{r_n} - k'^{s_n} m'^{r_n})$$

are positive integers.

For any $\varepsilon \in (0, 1/\mu)$ we can take a sufficiently large integer n_0 in such a way that (12.8), (12.11),

$$p_n > \frac{1}{\varepsilon} \cdot \frac{\log 2}{\log(m/m')},$$

$$s_{n-1} > \frac{1}{\varepsilon} \max\left(\mu, \frac{\log 2}{\log k + (1/\mu)\log m}\right),$$

$$\frac{s_n}{q_n} < \vartheta + \varepsilon \quad \text{and} \quad \frac{s_n}{s_{n-1}} < \vartheta^* + \varepsilon$$

hold for all $n \ge n_0$. Using $q_n < s_n$, $p_n < r_n$ and $r_n < (1/\mu + \varepsilon)s_n$ we have

$$1 \le \max(b_n, d_n) < k^{s_n+1} m^{r_n+1} < \sigma^{s_n},$$

where $\sigma = (km^{1/\mu+\varepsilon})^{1+\varepsilon}$. Moreover, using $q_n > s_n/(\vartheta + \varepsilon)$ and $p_n > (1/\mu - \varepsilon)q_n$, we obtain

$$2\left(\frac{k'}{k}\right)^{q_n+s_n}\left(\frac{m'}{m}\right)^{p_n+r_n} < \tau^{-s_n},$$

where

$$\tau = \left(\frac{k}{k'}\right)^{1+1/(\vartheta+\varepsilon)}\left(\frac{m}{m'}\right)^{(1-\varepsilon)(1/\mu+(1/\mu-\varepsilon)/(\vartheta+\varepsilon))} > 1.$$

This implies the irrationality of γ, because $\tau > \sigma$ for a sufficiently small ε by the condition (12.10), as is the case with the previous theorem.

Now for any integer q greater than

$$\max\left(2^{1/\varepsilon}, \frac{1}{2}\left(\frac{\tau}{\sigma}\right)^{s_{n_0}}\right),$$

we can define the integer $n > n_0$ uniquely by the inequalities

$$\left(\frac{\tau}{\sigma}\right)^{s_{n-1}} \le 2q < \left(\frac{\tau}{\sigma}\right)^{s_n},$$

because the sequence $\{s_n\}$ is strictly monotonically increasing. For any integer p we put $u_n/v_n = a_n/b_n$ if $a_n/b_n \ne p/q$; otherwise we put $u_n/v_n = c_n/d_n$. Hence, $u_n/v_n \ne p/q$, because $a_n/b_n \ne c_n/d_n$. Therefore it follows that

$$\begin{aligned}\left|\gamma - \frac{p}{q}\right| &\ge \left|\frac{p}{q} - \frac{u_n}{v_n}\right| - \left|\gamma - \frac{u_n}{v_n}\right| \\ &\ge \frac{1}{qv_n} - \frac{1}{\tau^{s_n}} \\ &> \frac{1}{q\sigma^{s_n}} - \frac{1}{\tau^{s_n}} \\ &> \frac{1}{2q\sigma^{s_n}}.\end{aligned}$$

Since $2 < \sigma^{\varepsilon s_n}$ and

$$\left(\frac{\tau}{\sigma}\right)^{s_n/(\vartheta^*+\varepsilon)} < \left(\frac{\tau}{\sigma}\right)^{s_{n-1}} \le 2q < q^{1+\varepsilon},$$

we obtain

$$\min_{p\in\mathbb{Z}} \frac{\log|\gamma - p/q|}{\log q} > -1 - (1+\varepsilon)^2 \frac{\vartheta^* + \varepsilon}{\xi - 1}$$

where $\xi = \log\tau/\log\sigma$. Taking the inferior limit as $q \to \infty$, we get

$$\omega(\gamma) \le 1 + (1+\varepsilon)^2 \frac{\vartheta^* + \varepsilon}{\xi - 1}$$
$$\le 1 + \vartheta\vartheta^* \left(1 - (\vartheta + 1)\frac{\mu\log k' + \log m'}{\mu\log k + \log m}\right)^{-1} + O(\varepsilon)$$

as $\varepsilon \to 0+$. Since ε is arbitrary, this completes the proof. □

Corollary 12.2. *Suppose that μ satisfies the same conditions in Theorem 12.4. Then for any integers $k, m \ge 2$, $M_\mu(1/k, 1/m)$ and $H_\mu(1/m)$ are irrational numbers with an irrationality measure $1 + \vartheta\vartheta^*$, and hence they are not Liouville numbers.*

12.4 Leaves outside First Quadrant

The structure of α- and β-leaves outside the first quadrant is not as simple as that in the first quadrant. The algebraic curves defined by

$$P^+_{p,q}(w,z) = 0 \quad \text{and} \quad P_{p,q}(w,z) = 0$$

represent the boundaries of $\alpha_{p/q}$- and $\beta_{p/q}$-leaves respectively. (See Exercise 4 in this chapter.) Fig. 12.1 shows the structural complexity of such algebraic curves. However, in the upper half-plane $z > 0$ we have the following

Lemma 12.3. *All $\alpha_{p/q}$- and $\beta_{p/q}$-leaves are smooth surfaces in the region*

$$\{(w,z) : w \ge 0, z > 0\} \cup \left\{(w,z) : -1 < w < 0 \text{ and } 0 < z < -\frac{1}{w}\right\}.$$

Proof. We assume that $-1 < w < 0, z > 0$ and $1 + wz > 0$. For any irreducible fraction p/q in the interval $(0,1)$ we have

$$P^+_{p,q}(w,z) = \sum_{k=1}^{q} w^{k-1} z^{\lceil kp/q \rceil - 1}$$
$$\ge \sum_{\ell=0}^{[q/2]-1} (1 + wz^{m(\ell)}) w^{2\ell} z^{n(\ell)},$$

where

$$m(\ell) = \left\lceil \frac{(2\ell+2)p}{q} \right\rceil - \left\lceil \frac{(2\ell+1)p}{q} \right\rceil \quad \text{and} \quad n(\ell) = \left\lceil \frac{(2\ell+1)p}{q} \right\rceil - 1.$$

Since $m(\ell) = 0$ or 1, we get

$$1 + wz^{m(\ell)} \ge 1 + w\max(1,z) > 0,$$

which implies that $P^+_{p,q}(w,z) > 0$. Similarly we have $P_{p,q}(w,z) > 0$. □

Unlike in the first quadrant, some leaves intersect mutually in the region

$$\mathfrak{D}_0 = \left\{(w,z) : -1 < w < 0 \text{ and } 0 < z < -\frac{1}{w}\right\}.$$

For example, it contains a part of the algebraic curve

$$1 + w + wz + w^2z - w^4z^2 = 0,$$

on which $\alpha_{1/3}$-leaf and $\beta_{2/3}$-leaf crossover.

The remainder of this section is devoted to studying the structure of specific leaves converging to the function

$$1 - \frac{Q_\varphi(1,w)}{wQ_{1/\varphi}(w,1)} = 1 - \left(\frac{1-w}{w}\right)^2 H_{1/\varphi}(w)$$

on the segment $\{-1 < w < 0, z = 1\}$ in $\mathfrak{D}_0$, where $\varphi = (\sqrt{5}+1)/2$ is the golden ratio.

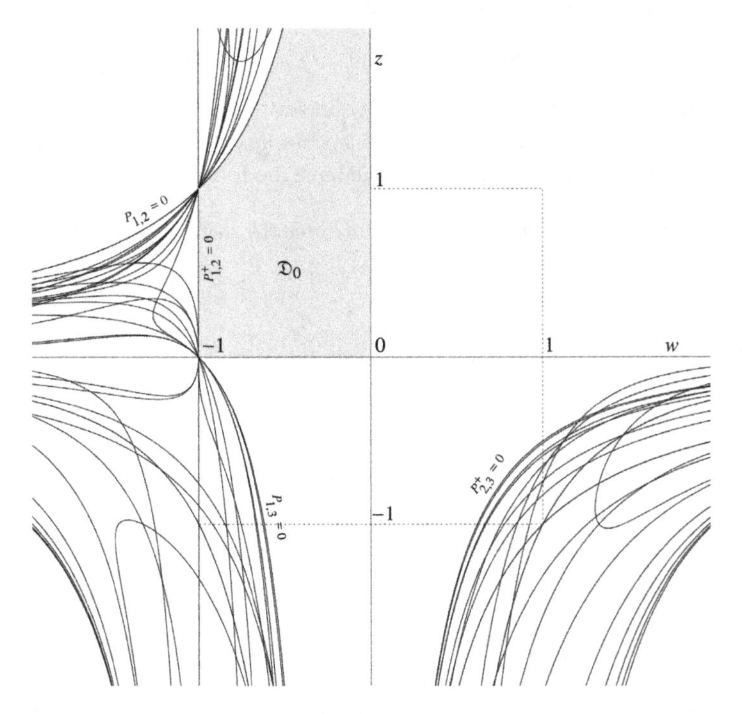

Fig. 12.1 The algebraic curves defined by $P^+_{p,q}(w,z) = 0$ and $P_{p,q}(w,z) = 0$ are plotted for all irreducible fractions p/q in the interval $(0,1)$ with $q \le 8$.

We first need a simple lemma concerning the Fibonacci numbers F_n.

Lemma 12.4. *For any positive integers m, n we have*

$$F_{n+1}F_{n+m} - F_nF_{n+m+1} = (-1)^n F_m.$$

Proof. For an arbitrarily fixed positive integer n, we put

$$a_m = (-1)^n(F_{n+1}F_{n+m} - F_nF_{n+m+1}).$$

Obviously $a_{m+2} = a_m + a_{m+1}$ holds for all $m \geq 1$. Since

$$a_1 = (-1)^n(F_{n+1}^2 - F_nF_{n+2}) = 1$$

and

$$\begin{aligned} a_2 &= (-1)^n(F_{n+1}F_{n+2} - F_nF_{n+3}) \\ &= (-1)^n(F_{n+1}^2 - F_nF_{n+2}) = 1, \end{aligned}$$

we get $a_m = F_m$ for any $m \geq 1$. □

Let $\{p_n/q_n\}$ and $\{r_n/s_n\}$ be the lower and the upper approximation sequences associated with $1/\varphi$ respectively. Since $\ell_n(1/\varphi) = 2$ for all $n \geq 1$, it can be seen that

$$\frac{p_n}{q_n} = \frac{F_{2n-2}}{F_{2n-1}} \quad \text{and} \quad \frac{r_n}{s_n} = \frac{F_{2n-1}}{F_{2n}}$$

for all $n \geq 2$. We write shortly

$$\alpha^{[n]}(w, z) = \alpha_{r_n/s_n}(w, z) \quad \text{and} \quad \beta^{[n]}(w, z) = \beta_{p_n/q_n}(w, z)$$

for $n \geq 1$. The following shows the monotonicity of the sequence $\alpha^{[3n-1]}(w, 1)$.

Lemma 12.5. *The rational functions $\alpha^{[3n-1]}(w, 1)$ form a strictly monotonically increasing sequence on the interval $-1 < w < 0$ and satisfy $\alpha^{[3n-1]}(-1, 1) = -1$ for all $n \geq 1$.*

Proof. Since

$$\begin{aligned} \alpha_{r/s}(w, 1) &= 1 - \frac{P_{s,r}(1, w)}{wP_{r,s}^{+}(w, 1)} \\ &= 1 - \frac{1-w}{1-w^s}\sum_{k=1}^{r} w^{[ks/r]-1}, \end{aligned}$$

the inequality $\alpha^{[3n+2]}(w,1) > \alpha^{[3n-1]}(w,1)$ is equivalent to

$$\sum_{k=1}^{F_{6n+3}} w^{[kF_{6n+4}/F_{6n+3}]} + \sum_{\ell=1}^{F_{6n-3}} w^{F_{6n+4}+[\ell F_{6n-2}/F_{6n-3}]} > \sum_{\ell=1}^{F_{6n-3}} w^{[\ell F_{6n-2}/F_{6n-3}]} + \sum_{k=1}^{F_{6n+3}} w^{F_{6n-2}+[kF_{6n+4}/F_{6n+3}]} \tag{12.12}$$

for $-1 < w < 0$. We now compare each term in the left-hand side of (12.12) with that in the right-hand side in this order, by distinguishing three cases as follows.

(a) The first F_{6n-3} terms.

Applying Lemma 12.4 with $m = 6$ we have

$$\left(\frac{F_{6n+4}}{F_{6n+3}} - \frac{F_{6n-2}}{F_{6n-3}}\right)k = \frac{8k}{F_{6n-3}F_{6n+3}} \in \left(0, \frac{1}{F_{6n-3}}\right)$$

for $1 \leq k \leq F_{6n-3}$, because $F_{6n+3} > 13F_{6n-3}$. Hence

$$\left[\frac{F_{6n+4}}{F_{6n+3}}k\right] = \left[\frac{F_{6n-2}}{F_{6n-3}}k\right],$$

which means that the first F_{6n-3} terms in the left-hand side of (12.12) coincide with that in the right-hand side.

(b) The last F_{6n-3} terms.

By the same reason as (a) we have

$$F_{6n+4} + \frac{F_{6n-2}}{F_{6n-3}}\ell - F_{6n-2} - \frac{F_{6n+4}}{F_{6n+3}}(\ell - F_{6n-3} + F_{6n+3}) = \frac{8}{F_{6n+3}}\left(1 - \frac{\ell}{F_{6n-3}}\right) \in \left[0, \frac{1}{F_{6n-3}}\right)$$

for $1 \leq \ell \leq F_{6n-3}$. Therefore,

$$F_{6n+4} + \left[\frac{F_{6n-2}}{F_{6n-3}}\ell\right] = F_{6n-2} + \left[\frac{F_{6n+4}}{F_{6n+3}}(\ell - F_{6n-3} + F_{6n+3})\right],$$

which implies that the last F_{6n-3} terms in the left-hand side of (12.12) coincide also with that in the right-hand side.

(c) The remainder terms.

For $F_{6n-3} < k \leq F_{6n+3}$ we obtain

$$\frac{F_{6n+4}}{F_{6n+3}}k - F_{6n-2} - \frac{F_{6n+4}}{F_{6n+3}}(k - F_{6n-3}) = \frac{8}{F_{6n+3}},$$

and hence,

$$\left[\frac{F_{6n+4}}{F_{6n+3}}k\right] = F_{6n-2} + \left[\frac{F_{6n+4}}{F_{6n+3}}(k - F_{6n-3})\right] + \delta_k,$$

where $\delta_k = 0$ or 1 according as the fractional part of kF_{6n+4}/F_{6n+3} is greater than $8/F_{6n+3}$ or not.

Summarizing the above arguments we conclude that the inequality (12.12) is equivalent to

$$f_n(w) = \sum_{\substack{F_{6n-3}<k\leq F_{6n+3} \\ \{kF_{6n+4}/F_{6n+3}\}<8/F_{6n+3}}} w^{L(k)} > 0$$

where $L(k) = [kF_{6n+4}/F_{6n+3}]$, because $1 - w^{-1} > 0$.

Our problem is thus reduced to the set of Diophantine equations

$$F_{6n+4}X - F_{6n+3}Y = i, \quad 0 \leq i \leq 7$$

satisfying $F_{6n-3} < X \leq F_{6n+3}$.[§] Lemma 12.4 for $m = 2$ and $m = 4$ gives the solutions $(X, Y) = (k, L(k))$ for $i = F_2 = 1$ and $i = F_4 = 3$; namely,

$$(X, Y) = (F_{6n+1}, F_{6n+2}) \quad \text{and} \quad (F_{6n-1}, F_{6n})$$

respectively. By combining these we get the solutions for $i = 2, 4, 5, 6$ and 7; that is,

$$\begin{aligned}
&(2F_{6n+1}, 2F_{6n+2}), \quad (F_{6n-1} + F_{6n+1}, F_{6n} + F_{6n+2}),\\
&(F_{6n-1} + 2F_{6n+1}, F_{6n} + 2F_{6n+2}),\\
&(2F_{6n-1}, 2F_{6n}) \quad \text{and} \quad (2F_{6n-1} + F_{6n+1}, 2F_{6n} + F_{6n+2})
\end{aligned}$$

respectively. Moreover $(X, Y) = (F_{6n+3}, F_{6n+4})$ is clearly the solution for $i = 0$. Therefore we have

$$f_n(w) = w^{F_{6n}}(1 + w^{F_{6n+1}})(1 + w^{F_{6n}} + w^{F_{6n}+F_{6n+2}} + w^{F_{6n+3}}),$$

which is positive for $-1 < w < 0$, because $1+w^\ell > 0$ for all $\ell \geq 1$ and F_k is an even integer if k is a multiple of 3. So the inequality (12.12) holds for $-1 < w < 0$.

Finally we have

$$\alpha^{[3n-1]}(-1, 1) = \alpha^{[2]}(-1, 1) = -1$$

for all $n \geq 1$, because $f_n(-1) = 0$. This completes the proof. □

[§] The solution (X, Y) of this equation is unique if exists. For otherwise, let (X', Y') be another solution with $X < X'$. Since $F_{6n+4}(X' - X) = F_{6n+3}(Y' - Y)$ and $\gcd(F_{6n+4}, F_{6n+3}) = 1$, we get $X' \equiv X \pmod{F_{6n+3}}$, and hence $X' > F_{6n+3}$, a contradiction.

In the above proof it is shown that

$$\alpha^{[3n+2]}(w,1)-\alpha^{[3n-1]}(w,1)=\left(\frac{1-w}{w}\right)^2\frac{f_n(w)}{(1-w^{F_{6n-2}})(1-w^{F_{6n+4}})}.$$

Thus we have by Lemma 8.2 the following

Theorem 12.5. *The expression*

$$H_{1/\varphi}(w)=\frac{w^2(1+w^2)}{(1-w)(1-w^3)}$$
$$-\sum_{n=1}^{\infty}\frac{w^{F_{6n}}(1+w^{F_{6n+1}})(1+w^{F_{6n}}+w^{F_{6n}+F_{6n+2}}+w^{F_{6n+3}})}{(1-w^{F_{6n-2}})(1-w^{F_{6n+4}})}$$

is valid for $|w|<1$.

The similar argument can be applied to the sequence $\beta^{[3n+1]}(w,1)$.

Lemma 12.6. *The rational functions* $\beta^{[3n+1]}(w,1)$ *form a strictly monotonically decreasing sequence on the interval* $-1<w<0$ *and satisfy* $\beta^{[3n+1]}(-1,1)=1$ *for all* $n\geq 1$.

Proof. Since

$$\beta_{p/q}(w,1)=1-\frac{P^{+}_{q,p}(1,w)}{wP_{p,q}(w,1)}$$
$$=1-\frac{1-w}{1-w^q}\sum_{k=1}^{p}w^{\lceil kq/p\rceil-2},$$

the inequality $\beta^{[3n+4]}(w,1)<\beta^{[3n+1]}(w,1)$ is equivalent to

$$\sum_{k=1}^{F_{6n+6}}w^{\lceil kF_{6n+7}/F_{6n+6}\rceil}+\sum_{\ell=1}^{F_{6n}}w^{F_{6n+7}+\lceil \ell F_{6n+1}/F_{6n}\rceil}$$
$$>\sum_{\ell=1}^{F_{6n}}w^{\lceil \ell F_{6n+1}/F_{6n}\rceil}+\sum_{k=1}^{F_{6n+6}}w^{F_{6n+1}+\lceil kF_{6n+7}/F_{6n+6}\rceil}\qquad(12.13)$$

for $-1<w<0$.

In the same way as the previous proof, we distinguish three cases as follows.

(a) The first F_{6n} terms.

We have

$$\left(\frac{F_{6n+1}}{F_{6n}}-\frac{F_{6n+7}}{F_{6n+6}}\right)\ell=\frac{8\ell}{F_{6n}F_{6n+6}}\in\left(0,\frac{1}{F_{6n}}\right)$$

for $1 \le \ell \le F_{6n}$, because $F_{6n+6} > 13F_{6n}$. Thus,

$$\left\lceil \frac{F_{6n+1}}{F_{6n}} \ell \right\rceil = \left\lceil \frac{F_{6n+7}}{F_{6n+6}} \ell \right\rceil,$$

which means that the first F_{6n} terms in the left-hand side of (12.13) coincide with that in the right-hand side.

(b) The last F_{6n} terms.

Similarly we have

$$\begin{aligned} F_{6n+1} + \frac{F_{6n+7}}{F_{6n+6}}(k - F_{6n} + F_{6n+6}) - F_{6n+7} - \frac{F_{6n+1}}{F_{6n}} k \\ = \frac{8}{F_{6n+6}}\left(1 - \frac{k}{F_{6n}}\right) \in \left[0, \frac{1}{F_{6n}}\right) \end{aligned}$$

for $1 \le k \le F_{6n}$. Therefore,

$$F_{6n+1} + \left\lceil \frac{F_{6n+7}}{F_{6n+6}}(k - F_{6n} + F_{6n+6}) \right\rceil = F_{6n+7} + \left\lceil \frac{F_{6n+1}}{F_{6n}} k \right\rceil,$$

so the last F_{6n} terms in the left-hand side of (12.13) coincide also with that in the right-hand side.

(c) The remainder terms.

For $F_{6n} < k \le F_{6n+6}$ it follows that

$$F_{6n+1} + \frac{F_{6n+7}}{F_{6n+6}}(k - F_{6n}) - \frac{F_{6n+7}}{F_{6n+6}} k = \frac{8}{F_{6n+6}},$$

and therefore

$$\left\lceil \frac{F_{6n+7}}{F_{6n+6}} k \right\rceil = F_{6n+1} + \left\lceil \frac{F_{6n+7}}{F_{6n+6}}(k - F_{6n}) \right\rceil - \delta_k,$$

where $\delta_k = 0$ or 1 according as the fractional part of kF_{6n+7}/F_{6n+6} is in the interval $[1/F_{6n+6}, 1 - 8/F_{6n+6}]$ or not.

From the above consideration we conclude that the inequality (12.13) is equivalent to

$$g_n(w) = \sum_{\substack{F_{6n} < k \le F_{6n+6} \\ \{kF_{6n+7}/F_{6n+6}\}=0 \text{ or } 1-i/F_{6n+6} \\ 1 \le i \le 7}} w^{L(k)} > 0,$$

where $L(k) = \lceil kF_{6n+7}/F_{6n+6} \rceil$, because $w - 1 < 0$. Clearly $\{kF_{6n+7}/F_{6n+6}\} = 0$ and $L(k) = F_{6n+7}$ when $k = F_{6n+6}$.

We now consider the set of Diophantine equations

$$F_{6n+7}X - F_{6n+6}Y = -i, \qquad 1 \le i \le 7$$

satisfying $F_{6n} < X \le F_{6n+6}$. Lemma 12.4 with $m = 2$ and $m = 4$ provides the solutions $(X, Y) = (k, L(k))$ for $i = 1$ and $i = 3$; namely,

$$(X, Y) = (F_{6n+4}, F_{6n+5}) \quad \text{and} \quad (F_{6n+2}, F_{6n+3})$$

respectively. Combining these we get the solutions for $i = 2, 4, 5, 6$ and 7; namely

$$\begin{aligned}&(2F_{6n+4}, 2F_{6n+5}), \quad (F_{6n+2} + F_{6n+4}, F_{6n+3} + F_{6n+5}),\\&(F_{6n+2} + 2F_{6n+4}, F_{6n+3} + 2F_{6n+5}),\\&(2F_{6n+2}, 2F_{6n+3}) \quad \text{and} \quad (2F_{6n+2} + F_{6n+4}, 2F_{6n+3} + F_{6n+5})\end{aligned}$$

respectively. Hence we obtain

$$g_n(w) = w^{F_{6n+3}}(1 + w^{F_{6n+4}})(1 + w^{F_{6n+3}} + w^{F_{6n+3}+F_{6n+5}} + w^{F_{6n+6}}),$$

which is positive for $-1 < w < 0$.

Finally we have $\beta^{[3n+1]}(-1, 1) = \beta^{[4]}(-1, 1) = 1$ for all $n \ge 1$, because $g_n(-1) = 0$. This completes the proof. □

In the above proof we show that

$$\beta^{[3n+4]}(w, 1) - \beta^{[3n+1]}(w, 1) = -\left(\frac{1-w}{w}\right)^2 \frac{g_n(w)}{(1 - w^{F_{6n+1}})(1 - w^{F_{6n+7}})},$$

which implies the following

Theorem 12.6. *The expression*

$$H_{1/\varphi}(w) = \sum_{n=0}^{\infty} \frac{w^{F_{6n+3}}(1 + w^{F_{6n+4}})(1 + w^{F_{6n+3}} + w^{F_{6n+3}+F_{6n+5}} + w^{F_{6n+6}})}{(1 - w^{F_{6n+1}})(1 - w^{F_{6n+7}})}$$

is valid for $|w| < 1$.

12.5 Infinite Oscillation of $H_{1/\varphi}$

As an application of Theorems 12.5 and 12.6 we show the following

Theorem 12.7. *The graph of the series $H_{1/\varphi}(w)$ oscillates infinitely often in the right-neighborhood of $w = -1$. More precisely, we have*

$$\limsup_{w \to -1+} H_{1/\varphi}(w) - \liminf_{w \to -1+} H_{1/\varphi}(w) > 0.06.$$

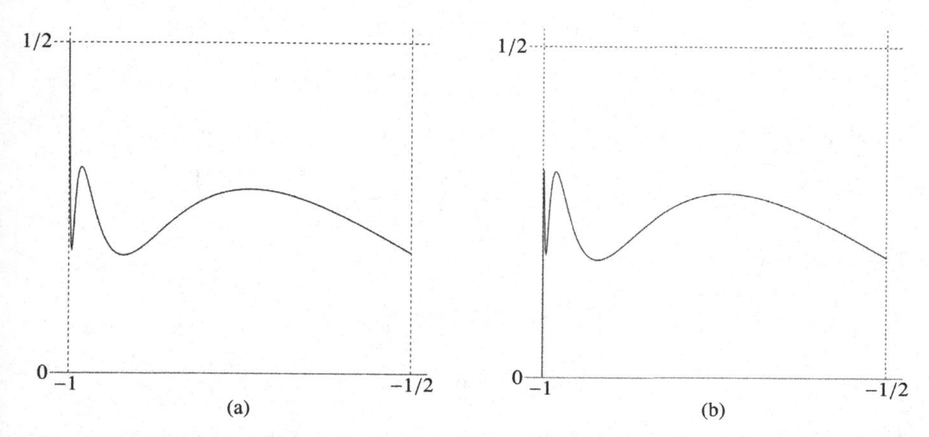

Fig. 12.2 (a) and (b) are the graphs of the truncated series of $H_{1/\varphi}(w)$, made of the first three terms of the series in Theorems 12.5 and 12.6 respectively. They are most of the same. Compare also with Fig. 8.1.

Consequently, the series

$$\frac{d}{dw}H_{1/\varphi}(w) = \sum_{n=1}^{\infty} n\left[\frac{n}{\varphi}\right]w^{n-1}$$

has infinitely many negative zeros.

Proof. For any $n \geq 1$ we put $A_n(w) = w^{F_{6n}}(1 + w^{F_{6n+1}})a_n(w)$, where

$$a_n(w) = \frac{1 + w^{F_{6n}} + w^{F_{6n}+F_{6n+2}} + w^{F_{6n+3}}}{(1 - w^{F_{6n-2}})(1 - w^{F_{6n+4}})}.$$

As is already seen, $A_n(w) > 0$ for $-1 < w < 0$. Let γ be a constant in the interval $(0, 1)$ to be determined later. We then define

$$x_n = -\gamma^{1/F_{6n-2}}$$

for $n \geq 1$, being a strictly monotonically decreasing sequence converging to -1 as $n \to \infty$. For any $\varepsilon \in (0, 1)$ take a sufficiently large integer n_0 satisfying

$$\left|\frac{F_{\ell+1}}{F_\ell} - \varphi\right| < \varepsilon \quad \text{and} \quad \left|\frac{F_\ell}{F_{\ell+1}} - \frac{1}{\varphi}\right| < \varepsilon$$

for all $\ell \geq 6n_0$. Since

$$a_n(x_n) > \frac{1 + \gamma^{(\varphi+\varepsilon)^2} + \gamma^{(\varphi+\varepsilon)^5} - \gamma^{(\varphi-\varepsilon)^2+(\varphi-\varepsilon)^4}}{(1+\gamma)(1+\gamma^{(\varphi-\varepsilon)^6})}$$

and

$$|x_n|^{F_{6n}}(1-|x_n|^{F_{6n+1}}) > \gamma^{(\varphi+\varepsilon)^2}(1-\gamma^{(\varphi-\varepsilon)^3})$$

for all $n > n_0$, we obtain $A_n(x_n) > \Phi(\gamma) + O(\varepsilon)$ as $\varepsilon \to 0+$, where

$$\Phi(\gamma) = \frac{(1-\gamma^{2\varphi+1})(\gamma^{\varphi+1}+\gamma^{2\varphi+2}+\gamma^{6\varphi+4}-\gamma^{5\varphi+4})}{(1+\gamma)(1+\gamma^{8\varphi+5})}.$$

Hence we have

$$\begin{aligned}\limsup_{w\to -1+}\sum_{k=1}^{\infty} A_k(w) &\geq \limsup_{n\to\infty}\sum_{k=1}^{\infty} A_k(x_n)\\ &\geq \limsup_{n\to\infty} A_n(x_n) \geq \Phi(\gamma),\end{aligned}$$

because ε is arbitrary. Thus it follows from Theorem 12.5 that

$$\begin{aligned}\liminf_{w\to -1+} H_{1/\varphi}(w) &= \frac{1}{2} - \limsup_{w\to -1+}\sum_{k=1}^{\infty} A_k(w)\\ &\leq \frac{1}{2} - \Phi(\gamma).\end{aligned} \tag{12.14}$$

On the other hand, we put $B_n(w) = w^{F_{6n+3}}(1+w^{F_{6n+4}})b_n(w)$, where

$$b_n(w) = \frac{1+w^{F_{6n+3}}+w^{F_{6n+3}+F_{6n+5}}+w^{F_{6n+6}}}{(1-w^{F_{6n+1}})(1-w^{F_{6n+7}})}$$

for any $n \geq 1$. Note that $B_n(w) > 0$ for any $-1 < w < 0$. We next define

$$y_n = -\gamma^{1/F_{6n+1}}$$

for $n \geq 1$, being a strictly monotonically decreasing sequence converging to -1 as $n \to \infty$. Since

$$b_n(y_n) > \frac{1+\gamma^{(\varphi+\varepsilon)^2}+\gamma^{(\varphi+\varepsilon)^5}-\gamma^{(\varphi-\varepsilon)^2+(\varphi-\varepsilon)^4}}{(1+\gamma)(1+\gamma^{(\varphi-\varepsilon)^6})}$$

and

$$|y_n|^{F_{6n+3}}(1-|y_n|^{F_{6n+4}}) > \gamma^{(\varphi+\varepsilon)^2}(1-\gamma^{(\varphi-\varepsilon)^3})$$

for all $n > n_0$, we obtain $B_n(y_n) > \Phi(\gamma) + O(\varepsilon)$ as $\varepsilon \to 0+$ in the same way as the previous case. Therefore it follows from Theorem 12.6 that

$$\begin{aligned}\limsup_{w\to -1+} H_{1/\varphi}(w) &\geq \limsup_{n\to\infty}\sum_{k=1}^{\infty} B_k(y_n)\\ &\geq \limsup_{n\to\infty} B_n(y_n) \geq \Phi(\gamma).\end{aligned} \tag{12.15}$$

Combining (12.14) with (12.15) we obtain

$$\limsup_{w\to -1+} H_{1/\varphi}(w) - \liminf_{w\to -1+} H_{1/\varphi}(w) \geq 2\Phi(\gamma) - \frac{1}{2}.$$

Now, it can be seen that $\Phi(0.8)$ is approximately 0.282889. This completes the proof. □

Theorem 12.7 means that α- and β-leaves are wrinkled in the neighborhood of the point $(-1, 1)$.

12.6 Sums involving Fibonacci Numbers

As another application of Theorems 12.5 and 12.6 we have the following formulae:

$$\sum_{n=1}^{\infty} \frac{1}{F_{6n-2}F_{6n+4}} = \frac{7-3\sqrt{5}}{48} \tag{12.16}$$

and

$$\sum_{n=1}^{\infty} \frac{1}{F_{6n-5}F_{6n+1}} = \frac{\sqrt{5}-1}{16}. \tag{12.17}$$

For the proof we need a lemma about the difference between $\alpha^{[3n-1]}(w, 1)$ and $\beta^{[3n+1]}(w, 1)$.

Lemma 12.7. *For all $n \geq 1$ we have*

$$\beta^{[3n+1]}(w, 1) - \alpha^{[3n-1]}(w, 1) = \left(\frac{1-w}{w}\right)^2 \frac{w^{F_{6n}} + w^{F_{6n-2}+F_{6n+1}}}{(1-w^{F_{6n-2}})(1-w^{F_{6n+1}})}.$$

Proof. The left-hand side of the expression in the lemma is, by definition, equal to

$$\frac{(1-w)(h_n^+(w) - h_n^-(w))}{(1-w^{F_{6n-2}})(1-w^{F_{6n+1}})},$$

where

$$h_n^+(w) = \sum_{k=1}^{F_{6n-3}} w^{[kF_{6n-2}/F_{6n-3}]-1} + \sum_{k=1}^{F_{6n}} w^{F_{6n-2}+\lceil kF_{6n+1}/F_{6n}\rceil-2},$$

$$h_n^-(w) = \sum_{k=1}^{F_{6n}} w^{\lceil kF_{6n+1}/F_{6n}\rceil-2} + \sum_{k=1}^{F_{6n-3}} w^{F_{6n+1}+[kF_{6n-2}/F_{6n-3}]-1}.$$

We now compare each term of $h_n^+(w)$ to that of $h_n^-(w)$ in this order, by distinguishing three cases as follows.

(a) The first F_{6n-3} terms.

Applying Lemma 12.4 with $m = 3$ we have

$$\left(\frac{F_{6n+1}}{F_{6n}} - \frac{F_{6n-2}}{F_{6n-3}}\right)k = \frac{2k}{F_{6n-3}F_{6n}} \in \left(0, \frac{1}{F_{6n-3}}\right)$$

for $1 \le k \le F_{6n-3}$. Hence the integral part of kF_{6n-2}/F_{6n-3} coincides with that of kF_{6n+1}/F_{6n}; so, the first F_{6n-3} terms in $h_n^+(w)$ and $h_n^-(w)$ cancel each other in the difference $h_n^+(w) - h_n^-(w)$, because $kF_{6n+1}/F_{6n} \notin \mathbb{Z}$.

(b) The last F_{6n-3} terms.

For $1 \le k < F_{6n-3}$ it follows that

$$\begin{aligned} F_{6n+1} &+ \frac{F_{6n-2}}{F_{6n-3}}k - F_{6n-2} - \frac{F_{6n+1}}{F_{6n}}(k + F_{6n} - F_{6n-3}) \\ &= \frac{2}{F_{6n}}\left(1 - \frac{k}{F_{6n-3}}\right) \in \left(0, \frac{1}{F_{6n-3}}\right). \end{aligned}$$

Since $kF_{6n-2}/F_{6n-3} \notin \mathbb{Z}$, we have

$$F_{6n+1} + \left\lfloor \frac{F_{6n-2}}{F_{6n-3}}k \right\rfloor - 1 = F_{6n-2} + \left\lceil \frac{F_{6n+1}}{F_{6n}}(k + F_{6n} - F_{6n-3}) \right\rceil - 2.$$

Thus the contribution in Case (b) to the difference $h_n^+(w) - h_n^-(w)$, which occurs only when $k = F_{6n-3}$, is $(1 - w)w^{F_{6n-2}+F_{6n+1}-2}$.

(c) The remainder terms.

For $1 \le k \le F_{6n} - F_{6n-3}$ we get

$$\frac{F_{6n+1}}{F_{6n}}(k + F_{6n-3}) - F_{6n-2} - \frac{F_{6n+1}}{F_{6n}}k = \frac{2}{F_{6n}},$$

and hence

$$\left\lceil \frac{F_{6n+1}}{F_{6n}}(k + F_{6n-3}) \right\rceil - 2 = F_{6n-2} + \left\lceil \frac{F_{6n+1}}{F_{6n}}k \right\rceil - 2 + \delta_k,$$

where $\delta_k = 0, 1$ and $\delta_k = 1$ if and only if $\{kF_{6n+1}/F_{6n}\} = 1 - 1/F_{6n}$, because $kF_{6n+1}/F_{6n} \notin \mathbb{Z}$. Applying Lemma 12.4 with $m = 2$, we see that

$$F_{6n-1}F_{6n} - F_{6n-2}F_{6n+1} = 1.$$

Therefore the contribution in Case (c) to the difference $h_n^+(w) - h_n^-(w)$, which occurs only when $k = F_{6n-2}$, is $(1 - w)w^{F_{6n}-2}$.

We thus conclude that

$$h_n^+(w) - h_n^-(w) = \frac{1-w}{w^2}(w^{F_{6n}} + w^{F_{6n-2}+F_{6n+1}}),$$

which completes the proof. □

We are now ready for showing the formula (12.16). For any integer $n \geq 1$ we put

$$s_n(w) = \frac{w^{F_{6n}}(1+w^{F_{6n+1}})(1+w^{F_{6n}}+w^{F_{6n}+F_{6n+2}}+w^{F_{6n+3}})}{(1+w+\cdots+w^{F_{6n-2}-1})(1+w+\cdots+w^{F_{6n+4}-1})}.$$

Then it follows from Theorem 12.5 that

$$\begin{aligned} S_m(w) &= \sum_{n>m} s_n(w) \\ &= \frac{w^2(1+w^2)}{1+w+w^2} - (1-w)^2 H_{1/\varphi}(w) - \sum_{n=1}^{m} s_n(w). \end{aligned}$$

It is easily seen that the right-hand side converges as $w \to 1-$ to

$$\lim_{w\to 1-} S_m(w) = \frac{2}{3} - \frac{1}{\varphi} - \sum_{n=1}^{m} \frac{8}{F_{6n-2}F_{6n+4}}.$$

On the other hand, we have

$$\begin{aligned} 0 < \frac{S_m(w)}{w^2} &= \sum_{n>m} (\alpha^{[3n+2]}(w,1) - \alpha^{[3n-1]}(w,1)) \\ &< \beta^{[3m+4]}(w,1) - \alpha^{[3m+2]}(w,1) \end{aligned}$$

for any $0 < w < 1$, and hence by Lemma 12.7,

$$0 \leq \lim_{w\to 1-} S_m(w) \leq \frac{2}{F_{6m+4}F_{6m+7}}.$$

Thus we get

$$\lim_{m\to\infty} \lim_{w\to 1-} S_m(w) = 0,$$

which implies (12.16). The formula (12.17) can be shown in the same manner.

Exercises in Chapter 12

1 Let $\{q_n\}$ and $\{s_n\}$ be the denominators of the lower and the upper approximation sequences associated with an irrational κ respectively. Show that

$$\left(\liminf_{n\to\infty} \frac{s_n}{q_n}\right)^{-1} + \left(\limsup_{n\to\infty} \frac{s_{n+1}}{s_n}\right)^{-1} = 1.$$

2 Let $m \geq 2$ be an integer and μ be an irrational number having an irrationality measure $\omega \geq 2$. Show that $m\omega$ is an irrationality measure of $\mu^{1/m}$.

3 Show the following formula for the Fibonacci number:

$$F_n = \sum_{k=0}^{[(n-1)/2]} \binom{n-k-1}{k}.$$

4 Let p, q be coprime positive integers. Show that the polynomials $P_{p,q}(w, z)$ and $P^+_{q,p}(z, w)$ do not vanish simultaneously for any $(w, z) \in \mathbb{R}^2$.

5 Show that all $\alpha_{p/q}$- and $\beta_{p/q}$-leaves are smooth surfaces in the region defined by

$$\mathfrak{D}_1 = \left\{(w, z) : w < -1 \text{ and } z > -\frac{1}{w}\right\}$$

in the second quadrant.

6 Show that all $\alpha_{p/q}$- and $\beta_{p/q}$-leaves are smooth surfaces in the region defined by

$$\mathfrak{D}_2 = \left\{(w, z) : |w| \leq \frac{1}{2} \text{ and } -\frac{1}{2|w|} \leq z \leq 0\right\}$$

in the third and fourth quadrants.

Hints and Solutions

Chapter 1

1. Use the intermediate value theorem to $f(x) - x$.

2. For any initial value $x_0 \in X$ put $x_n = f^n(x_0)$. Then, for all positive integer n,

$$d(x_{n+1}, x_n) \le \lambda d(x_n, x_{n-1}) \le \cdots \le \lambda^n d(x_1, x_0),$$

which implies that

$$d(x_m, x_n) \le \sum_{k=n}^{m-1} d(x_{k+1}, x_k) \le d(x_1, x_0) \sum_{k=n}^{m-1} \lambda^k < \frac{d(x_1, x_0)}{1-\lambda} \lambda^n$$

for all integer m greater than n. Hence $\{x_n\}_{n \ge 0}$ is a Cauchy sequence and so it converges to some $x^* \in X$. Obviously we have $x^* \in \mathrm{Fix}(f)$. Suppose that there exist two fixed points of f, say x^* and x^{**}. Since

$$d(x^*, x^{**}) = d(f(x^*), f(x^{**})) \le \lambda d(x^*, x^{**}),$$

we have $d(x^*, x^{**}) = 0$; so $x^* = x^{**}$. Thus every orbit is drawn into this fixed point.

A weaker condition that $d(f(x), f(y)) < d(x, y)$ for any $x, y \in X$ is not sufficient to ensure the existence of a fixed point of f in general. However it is still true that f has a unique fixed point, if X is a compact metric space and $d(f(x), f(y)) < d(x, y)$ holds for any $x, y \in X$.

3. Suppose that $x \in I$ satisfies $S(f^n)(x) < 0$ and $(f^n)'(x) \ne 0$ for some $n \ge 1$. If

$$(f^{n+1})'(x) = \prod_{k=0}^{n} f'(f^k(x)) \ne 0,$$

then $(f^n)'(f(x)) \ne 0$ and $(f^n)'(x) \ne 0$; hence we have

$$S(f^{n+1})(x) = S(f^n)(f(x)) \cdot (f'(x))^2 + S(f^n)(x) < 0.$$

4. If $mx \notin \mathbb{Z}$, then

$$\tilde{\psi}(1-x) = \lim_{t \to (1-x)-} \{mt\} = \{-mx\} = 1 - \{mx\}.$$

On the other hand, for $0 \le i < m$ we have

$$\tilde{\psi}\left(1 - \frac{i}{m}\right) = \lim_{t \to (1-i/m)-} \{mt\} = 1 = 1 - \left\{\frac{i}{m} m\right\}.$$

[5] It can be seen that

$$h \circ \phi(x) = \sin^2\left(\frac{\pi}{2}(1 - |1 - 2x|)\right) = \sin^2(\pi x)$$

and

$$f \circ h(x) = 4\sin^2\frac{\pi x}{2}\left(1 - \sin^2\frac{\pi x}{2}\right) = \sin^2(\pi x).$$

[6] Recall that a function $d : X \times X \to \mathbb{R}$ is a metric on a set X provided that

(i) $d(x, y) \ge 0$ and $d(x, y) = 0$ if and only if $x = y$,
(ii) $d(x, y) = d(y, x)$,
(iii) $d(x, z) \le d(x, y) + d(y, z)$.

It is straightforward to verify that (1.4) satisfies the above axioms. In metric spaces "compactness" and "sequential compactness" are equivalent properties, and one can easily confirm that any infinite sequence in Σ contains a converging subsequence.

[7] Consider the following sequence:

$$\underbrace{0\;1}_{\text{1st block}}\;\underbrace{00\,01\,10\,11}_{\text{2nd block}}\;\underbrace{000\,001\,010\,011\,100\,101\,110\,111}_{\text{3rd block}}\cdots,$$

where n-th block consists of 2^n words of length n lining up in their natural order; so, this infinite sequence contains every possible finite word somewhere. The iterations of the shift operator σ pull out every word from the sequence, just like a jack-in-the-box.

[8] For an arbitrarily fixed $x \in X$ suppose first that $\mathrm{Orb}_f(x)$ is finite. Define the mapping $h : \mathbb{N} \cup \{0\} \to X$ by $h(n) = f^n(x)$. The mapping h is not one-to-one; for otherwise, $\mathrm{Orb}_f(x)$ would be infinite. Thus there exist two integers $0 \le m < n$ satisfying $f^m(x) = f^n(x)$; therefore

$$f^{n-m} \circ f^m(x) = f^n(x) = f^m(x)$$

and hence $f^m(x) \in \mathrm{Per}_d(f)$ for some divisor d of $n - m$.

Suppose conversely that $f^m(x) \in \mathrm{Per}_n(f)$ for some $m \ge 0, n \ge 1$. Then obviously $\mathrm{Orb}_f(x)$ is finite, because

$$\mathrm{Orb}_f(x) \subset \left\{x, f(x), \ldots, f^m(x), \ldots, f^{m+n-1}(x)\right\}.$$

[9] Since it is trivial for the case $\{a_n\} = \{b_n\}$, we can assume that there exists an integer $k \ge 1$ satisfying $\{a_n\} = a_1 a_2 \cdots a_k 0 \cdots$ and $\{b_n\} = a_1 a_2 \cdots a_k 1 \cdots$. Then

$$d(\{a_n\}, \{b_n\}) \ge \frac{1}{2^{k+1}}.$$

Moreover, since

$$\tau(\{a_n\}) = \varsigma(a_1)\cdots\varsigma(a_k)01\cdots \quad \text{and} \quad \tau(\{b_n\}) = \varsigma(a_1)\cdots\varsigma(a_k)0\cdots,$$

it follows that

$$d(\tau(\{a_n\}), \tau(\{b_n\})) \leq \frac{1}{2^{k+2}} + \frac{1}{2^{k+3}} + \cdots = \frac{1}{2^{k+1}},$$

which implies the required inequality.

On the other hand, if $\{a_n\} = 000\cdots$ and $\{b_n\} = 100\cdots$, then $\tau(\{a_n\}) = 010101\cdots$ and $\tau(\{b_n\}) = 00101\cdots$; hence

$$d(\{a_n\}, \{b_n\}) = \frac{1}{2} = d(\tau(\{a_n\}), \tau(\{b_n\})),$$

which means that τ is not a contraction.

Chapter 3

1 We prove the proposition by induction on n. For any $x \in (0, 1] \setminus \{c_0\}$ we have

$$f^*(1-x) = 1 - \tilde{f}(x) = 1 - f(x)$$

and this is also true at $x = 0$ by continuity. Suppose next that the proposition is true for $n = m$ for some positive integer m. For any $x \in [0, 1] \setminus \{c_0, \ldots, c_{-m}\}$ it follows that

$$\begin{aligned} f^{*m+1}(1-x) &= f^* \circ f^{*m}(1-x) = f^*(1 - f^m(x)) \\ &= 1 - f \circ f^m(x) = 1 - f^{m+1}(x); \end{aligned}$$

so, the proposition is also true for $n = m + 1$.

2 By definition

$$\psi^*_{a,b,c}(x) = 1 - \tilde{\psi}_{a,b,c}(1-x) = \begin{cases} a(x - 1 + c) & (0 \leq 1 - x \leq c), \\ 1 + b(x - 1 + c) & (c < 1 - x \leq 1). \end{cases}$$

3 Taking positive constants a, b with $ab < 1$ and $c = b/(1 + b)$, we have

$$\psi^2_{a,b,c}(x) = \begin{cases} c + ab(x - c) & (0 \leq x < c), \\ 1 + ab(x - 1) & (c \leq x \leq 1). \end{cases}$$

Then $\psi^{2n}_{a,b,c}(0)$ and $\psi^{2n+1}_{a,b,c}(0)$ are strictly monotonically increasing sequences converging to c and 1 respectively.

4 By Lemma 1.1 the set $[0, 1] \setminus \mathrm{Orb}_f(c)$ is not empty. Note that $x \in (0, c) \setminus \mathrm{Orb}_f(c)$ if and only if $h(x) = 1 - x \in (1 - c, 1) \setminus \mathrm{Orb}_f(c)$, where $1 - c$ is the discontinuity point of the reflection f^*.

5 For any integers $k \geq 2, n \geq 1$ we put $n = qk + r$ with $q \geq 0, 1 \leq r \leq k$. Applying the given inequality q times we have $a_n \leq qa_k + a_r$; hence,

$$\frac{a_n}{n} \leq \frac{n-r}{n} \cdot \frac{a_k}{k} + \frac{a_r}{n}.$$

Therefore

$$\limsup_{n\to\infty} \frac{a_n}{n} \leq \frac{a_k}{k}$$

for all $k \geq 1$; so, we have

$$\limsup_{n\to\infty} \frac{a_n}{n} \leq \inf_{k\geq 1} \frac{a_k}{k} \leq \liminf_{n\to\infty} \frac{a_n}{n}.$$

This implies (i) and (ii).

For any $x \in [0,1]$ and $y > 0$ it follows from (ii) and (iv) of Lemma 3.2 that

$$\begin{aligned} |F^n(x) - F^n(y)| &= \left|F^n(x) - F^n([y] + \{y\})\right| \\ &\leq \left|F^n(x) - F^n(\{y\})\right| + [y] \leq 1 + y. \end{aligned}$$

Hence, putting $y = F^m(x)$ we have

$$F^{m+n}(x) \leq F^m(x) + F^n(x) + 1$$

for all $m, n \geq 1$. Since the sequence $a_n = F^n(x) + 1$ satisfies $a_{m+n} \leq a_m + a_n$, we see from (i) that $F^n(x)/n$ converges as $n \to \infty$ for any $0 \leq x \leq 1$. This limit is independent of the choice of x by (ii) and (iv) of Lemma 3.2.

6 Let $c = c_0, \ldots, c_{1-m}$ $(m \geq 1)$ be all different inverse images of c in the interval $(0,1)$. It is easily seen that $\epsilon_n(c_{-k}+) = \epsilon_n(c_{-k}-)$ for $0 \leq n < k$ and that $\epsilon_n(c_{-k}\pm) = \epsilon_{n-k}(c\pm)$ for $n \geq k$ respectively; hence

$$\Theta_f(c_{-k}+, t) - \Theta_f(c_{-k}-, t) = t^k(\Theta_f(c+, t) - \Theta_f(c-, t)).$$

On the other hand, since $\epsilon_n(c+) = \epsilon_{n-1}(0)$ and $\epsilon_n(c-) = \epsilon_{n-1}(1-)$ for $n \geq 1$, it follows from (3.4) that

$$\begin{aligned} \Theta_f(c+, t) - \Theta_f(c-, t) &= 1 - t(\Theta_f(1-, t) - \Theta_f(0, t)) \\ &= 1 - t\sum_{k=0}^{m-1}(\Theta_f(c_{-k}+, t) - \Theta_f(c_{-k}-, t)) \\ &= 1 - \frac{t(1-t^m)}{1-t}(\Theta_f(c+, t) - \Theta_f(c-, t)). \end{aligned}$$

If $m = \infty$, then t^m should be replaced by 0.

7 Let $a_n = (-1)^n n$ for $n \geq 0$. It can be easily seen that $s_n = (-1)^n[(n+1)/2]$ and that

$$\frac{s_0 + s_1 + \cdots + s_n}{n} = \begin{cases} 0 & (n : \text{even}), \\ -\dfrac{n+1}{2n} & (n : \text{odd}). \end{cases}$$

On the other hand, we have

$$f(x) = \sum_{n=0}^{\infty}(-1)^n n x^n = -\frac{x}{(1+x)^2},$$

which converges to $-1/4$ as $x \to 1-$.

Chapter 4

1 Let $z_1 < z_2 < \cdots < z_q$ be the periodic points, numbered in their natural order. Then the permutation π of $\{1, 2, \ldots, q\}$ is induced by $f(z_k) = z_{\pi(k)}$. Apply the similar argument given in the proof of Theorem 4.2.

2 Put $[kp/q] = m-1$ and $[(k+1)p/q] = m$. We have $1 \le m \le p$ and $mq/p - 1 \le k < mq/p$. Therefore, if $1 \le m < p$, then $mq/p \notin \mathbb{Z}$ and so $mq/p - 1 < k < mq/p$; hence $k = [mq/p]$. If $m = p$, then clearly $k = q - 1$. Note that $Q_{p,q}(t)$ can be expressed simply as

$$Q_{p,q}(t) = \sum_{m=1}^{p} t^{\lceil mq/p \rceil - 1},$$

where $\lceil x \rceil$ is the ceiling function soon-to-be explained in Section 5.4.

3 By the expression of $Q_{p,q}$ stated in the previous exercise, the formula to be demonstrated is equivalent to

$$\sum_{m=1}^{p+r-1} t^{[m(q+s)/(p+r)]} = \sum_{m=1}^{p-1} t^{[mq/p]} + t^{q-1} + \sum_{m=1}^{r-1} t^{q+[ms/r]}. \tag{1}$$

To see this we distinguish three cases as follows.

(a) The first $p - 1$ terms.

Since

$$0 < \left(\frac{q}{p} - \frac{q+s}{p+r}\right)m = \frac{m}{p(p+r)} < \frac{1}{p+r}$$

for $1 \le m < p$, we have $[m(q+s)/(p+r)] = [mq/p]$; that is, the first $p - 1$ terms in the left-hand side of (1) coincide with that in the right-hand side.

(b) The p-th term.

Since $p(q+s)/(p+r) = q - 1/(p+r)$, we have $[p(q+s)/(p+r)] = q - 1$.

(c) The last $r - 1$ terms.

For $1 \le m < r$ we have

$$0 < \frac{s}{r}m + q - \frac{q+s}{p+r}(m+p) = \frac{r-m}{r(p+r)} < \frac{1}{p+r};$$

thus, $[(m+p)(q+s)/(p+r)] = q + [ms/r]$; that is, the last $r-1$ terms in the left-hand side of (1) coincide with that in the right-hand side.

[4] By definition we have

$$Q'_{p,q}(1) = \sum_{\substack{1 \le k < q \\ \{kp/q\}+p/q \ge 1}} k.$$

On the other hand, it follows from the formula in Exercise [2] in this chapter that

$$\begin{aligned} Q'_{p,q}(1) &= \sum_{m=1}^{p-1} \left[\frac{q}{p} m \right] + q - 1 \\ &= \frac{q}{p} \sum_{m=1}^{p-1} m - \left(\frac{1}{p} + \cdots + \frac{p-1}{p} \right) + q - 1 \\ &= \frac{(p+1)(q-1)}{2}. \end{aligned}$$

[5] Suppose first that $f \in \mathfrak{E}^0(p/q)$. By (ii) of Theorem 4.7 we get $0 \in \mathrm{Per}_q(f)$. Then we can apply the formula in Exercise [1] in Chapter 3 for $x = 0$ and $n = q-1$, because $c_0, \ldots, c_{2-q}$ are all positive. Thus,

$$f^{*q-1}(1) = 1 - f^{q-1}(0) = 1 - c \quad \text{and} \quad f^{*q}(1) = f^*(1-c) = 1 - \tilde{f}(c) = 0.$$

Therefore $f^* \in \mathfrak{E}^\phi(1 - p/q)$, because

$$\begin{aligned} \#\left\{0 \le k < q : f^{*k}(0) \ge 1 - c\right\} &= \#\left\{0 \le k < q : f^k(1) \le c\right\} \\ &= q - \#\left\{0 \le k < q : f^k(1) > c\right\} \\ &= q - p. \end{aligned}$$

We next suppose that $f^* \in \mathfrak{E}^\phi(1 - p/q)$. Since $f^{*q}(1) = 0$ by (i) of Theorem 4.7 we have $f^{*q-1}(1) = 1 - c$. So, if $c_{-1}, \ldots, c_{2-q} > 0$, then, applying the formula in Exercise [1] in Chapter 3 for $x = 0$ and $n = q-1$, we obtain $f^{q-1}(0) = 1 - f^{*q-1}(1) = c$ and so $f^q(0) = 0$. If $0 \in \mathrm{Per}_\ell(f)$ for some $\ell < q$, then the argument as described above implies that $f^{*\ell}(1) = 0$, a contradiction. Thus we have $0 \in \mathrm{Per}_q(f)$.

However, if $c_{-i} = 0$ for some integer $1 \le i \le 2 - q$, then 0 would be a periodic point with period $\ell < q$, again a contradiction. This completes the proof.

Chapter 5

[1] Suppose, on the contrary, that $q + s \le n$. Then the mediant $(p+r)/(q+s)$ would belong to $\mathcal{F}_n$, a contradiction.

[2] $p/q \in \mathcal{F}_n$ if and only if $(q-p)/q \in \mathcal{F}_n$.

[3] Put $qr - ps = \ell$. Any two contiguous fractions in $\mathcal{G}_n$, say $a/b < c/d$, satisfy $bc - ad = \ell$ and $b + d \ge n + 1$. Thus, for any irreducible fraction x/y in the interval $(a/b, c/d)$ we have

$$\frac{\ell}{bd} = \frac{c}{d} - \frac{a}{b} = \frac{c}{d} - \frac{x}{y} + \frac{x}{y} - \frac{a}{b} \ge \frac{1}{dy} + \frac{1}{by}.$$

Therefore $\ell y \ge b + d \ge n + 1$. This implies that any irreducible fraction x/y in the interval $(p/q, r/s)$ must coincide with the reduced fraction of some member in $\mathcal{G}_{\ell y}$.

[4] Since $\langle 1, 2, \ldots, n - 1 \rangle = [1 - 1/n, 1)$, we have

$$\bigcap_{n=2}^{\infty} \overline{\langle 1, 2, 3, \ldots, n \rangle} = 1.$$

[5] If κ is rational, then there exists a positive integer n_0 satisfying $q_n = q$ and $\ell_n(\kappa) = 1$ for all $n \ge n_0$. Thus $s_{n_0+k} = kq + s_{n_0}$ for all $k \ge 0$ and s_{n+1}/s_n tends to 1 as $n \to \infty$. Conversely, suppose that s_{n+1}/s_n tends to 1 as $n \to \infty$. Since

$$\ell_n(\kappa) = \frac{s_{n+1}}{s_n} + \frac{s_{n-1}}{s_n} - 1,$$

we have $\ell_n(\kappa) = 1$ for all sufficiently large n; so, κ must be rational.

[6] Since κ and r_{n+1}/s_{n+1} belong to the same Farey interval $[p_n/q_n, r_n/s_n)$, it is obvious that $\ell_k(\kappa) = \ell_k(r_{n+1}/s_{n+1})$ for $1 \le k < n$. We also have $\ell_n(\kappa) + 1 = \ell_n(r_{n+1}/s_{n+1})$ by definition; hence the desired expression for r_{n+1}/s_{n+1} follows from the recurrence relation (5.6) using

$$\phi^n\left(\frac{r_{n+1}}{s_{n+1}}\right) = 0.$$

Note that $p^*_{n+1} = r_{n+1}$ and $q^*_{n+1} = s_{n+1}$ where p^*_k/q^*_k is the lower approximation sequence associated with r_{n+1}/s_{n+1}. Replacing the last denominator $\ell_n + 1$ by ℓ_n we get the similar expression for p_{n+1}/q_{n+1}.

[7] The values of $3 + \dfrac{r_n}{s_n}$ for $n = 7, 8, 9$ are $\dfrac{22}{7}, \dfrac{355}{113}$ and $\dfrac{104348}{33215}$ respectively. Note that

$$\frac{22}{7} - \pi = 7^{-3.429\ldots}, \quad \frac{355}{113} - \pi = 113^{-3.201\ldots} \text{ and } \frac{104348}{33215} - \pi = 33215^{-2.096\ldots}.$$

[8] In the same way as the proof of Lemma 5.11 we see that

$$\begin{aligned}\frac{s_{n-1}}{s_{n-2}} &= [\ell_{n-2} + 1, \ldots, \ell_1 + 1] \\ &\ge [\underbrace{2, \ldots, 2}_{d \text{ times}}, 3, \ldots, \underbrace{2, \ldots, 2}_{d \text{ times}}, 3, [\underbrace{2, \ldots, 2}_{m \text{ times}}]],\end{aligned}$$

where the string "2, ..., 2, 3" of length $d + 1$ repeats k times and $m = n - (d + 1)k - 2$

is chosen so that $1 \le m \le d+1$. The rational function

$$g(x) = [\underbrace{2, \ldots, 2}_{d \text{ times}}, 3, x] = \frac{(2d+3)x - d - 1}{(2d+1)x - d}$$

maps the interval $[1, \infty)$ into itself and possesses a unique fixed point $x_d > 1$, which satisfies the quadratic equation $(2d+1)x^2 - 3(d+1)x + d + 1 = 0$; that is,

$$x_d = \frac{3(d+1) + \sqrt{(d+1)(d+5)}}{2(2d+1)}.$$

Then

$$\frac{s_{n-1}}{s_{n-2}} \ge g^k\left(1 + \frac{1}{m}\right)$$

and the right-hand side converges to x_d as $k \to \infty$, because x_d is a stable fixed point of g. Therefore, for any $\varepsilon > 0$ we have

$$\frac{s_n}{q_n} = 1 + \frac{1}{\ell_{n-1} - s_{n-2}/s_{n-1}} \le 1 + \frac{1}{2 - (x_d - \varepsilon)^{-1}}$$

for all sufficiently large n. Note that $x_0 = \varpi^{-1}$ formally, where ϖ is the constant stated in Lemma 5.10.

9 Suppose first that $\phi^m(\kappa)$ is a non-zero periodic point of ϕ with period $k \ge 1$. Since $\phi^m(\kappa) = \phi^{k+m}(\kappa)$, it follows from Lemma 5.12 that $a\kappa^2 - b\kappa + c = 0$, where

$$\begin{aligned} a &= q_{k+m+1}s_m - q_{m+1}s_{k+m}, \\ b &= q_{k+m+1}r_m - q_{m+1}r_{k+m} + p_{k+m+1}s_m - p_{m+1}s_{k+m}, \\ c &= p_{k+m+1}r_m - p_{m+1}r_{k+m}. \end{aligned}$$

If $a = 0$, then s_m would be a multiple of s_{k+m} because $\gcd(q_{n+1}, s_n) = 1$ for all $n \ge 1$, a contradiction. Hence $a \ne 0$ and κ is a quadratic irrational, because the orbit of κ does not fall into 0.

Suppose, conversely, that κ is a quadratic irrational; namely, $A\kappa^2 + B\kappa + C = 0$ for some integers $A \ge 1$, B and C. We put $\kappa_n = \phi^n(\kappa)$ and $P(x, y) = Axy + Bx + C$ for brevity. By Lemma 5.12 we get

$$\kappa = \frac{\kappa_n r_n + p_{n+1}}{\kappa_n s_n + q_{n+1}};$$

thus $A_n\kappa_n^2 + B_n\kappa_n + C_n = 0$ for $n \ge 1$, where

$$\begin{aligned} A_n &= Ar_n^2 + Br_ns_n + Cs_n^2 = s_n^2 P\left(\frac{r_n}{s_n}, \frac{r_n}{s_n}\right), \\ B_n &= 2Ap_{n+1}r_n + B(q_{n+1}r_n + p_{n+1}s_n) + 2Cq_{n+1}s_n \\ &= q_{n+1}s_n\left(P\left(\frac{r_n}{s_n}, \frac{p_{n+1}}{q_{n+1}}\right) + P\left(\frac{p_{n+1}}{q_{n+1}}, \frac{r_n}{s_n}\right)\right), \\ C_n &= Ap_{n+1}^2 + Bp_{n+1}q_{n+1} + Cq_{n+1}^2 = q_{n+1}^2 P\left(\frac{p_{n+1}}{q_{n+1}}, \frac{p_{n+1}}{q_{n+1}}\right). \end{aligned}$$

If $A_n = 0$, then r_n/s_n would be a conjugated root of κ, a contradiction. Hence $A_n \neq 0$ for all $n \geq 1$. Similarly $C_n \neq 0$ for all $n \geq 1$. Moreover, if

$$\begin{pmatrix} X \\ Y \\ Z \end{pmatrix} = \begin{pmatrix} s^2 & st & t^2 \\ 2su & sv+tu & 2tv \\ u^2 & uv & v^2 \end{pmatrix} \begin{pmatrix} x \\ y \\ z \end{pmatrix}, \tag{2}$$

then it can be easily seen that†

$$Y^2 - 4XZ = (sv - tu)^2(y^2 - 4xz).$$

(Note that the determinant of the matrix of size 3 in (2) is $(sv - tu)^3$.) Substituting $A, B, C, A_n, B_n, C_n, r_n, s_n, p_{n+1}, q_{n+1}$ for $x, y, z, X, Y, Z, s, t, u, v$ respectively, we get

$$B_n^2 - 4A_nC_n = (q_{n+1}r_n - p_{n+1}s_n)^2(B^2 - 4AC) = B^2 - 4AC, \tag{3}$$

because

$$\begin{aligned} q_{n+1}r_n - p_{n+1}s_n &= (q_n + (\ell_n - 1)s_n)r_n - (p_n + (\ell_n - 1)r_n)s_n \\ &= q_nr_n - p_ns_n = 1. \end{aligned}$$

On the other hand, put $\eta = \kappa - p_{n+1}/q_{n+1}$ and $\theta = r_n/s_n - \kappa$ for brevity. Since

$$P(x,y) = (A\kappa + B)(x - \kappa) + A\kappa(y - \kappa) + A(x - \kappa)(y - \kappa),$$

we have

$$\begin{aligned} |B_n| &\leq q_{n+1}s_n((2A\kappa + |B|)(\eta + \theta) + 2A\eta\theta), \\ |C_n| &\leq q_{n+1}^2((2A\kappa + |B|)\eta + A\eta^2). \end{aligned}$$

Hence it follows from

$$\eta < \frac{r_{n+1}}{s_{n+1}} - \frac{p_{n+1}}{q_{n+1}} = \frac{1}{q_{n+1}s_{n+1}} \quad \text{and} \quad \theta < \frac{r_n}{s_n} - \frac{p_{n+1}}{q_{n+1}} = \frac{1}{q_{n+1}s_n}$$

that both B_n and C_n are bounded sequences of integers, as well as A_n, because by (3)

$$|A_n| = \frac{|B_n^2 - B^2 + 4AC|}{4|C_n|} \leq \frac{1}{4}(B_n^2 + |B^2 - 4AC|).$$

Therefore we conclude that there exist three positive integers $n_1 < n_2 < n_3$ in such a way that $\kappa_{n_1}, \kappa_{n_2}$ and κ_{n_3} satisfy the same quadratic equation. Thus at least two of them must be equal.

---

† Prof. S. Kato at Kyoto University kindly informed the author that this identity comes easily from taking the determinants of the following alternative expression for the transformation (2):

$$\begin{pmatrix} X & Y/2 \\ Y/2 & Z \end{pmatrix} = \begin{pmatrix} s & t \\ u & v \end{pmatrix} \begin{pmatrix} x & y/2 \\ y/2 & z \end{pmatrix} \begin{pmatrix} s & u \\ t & v \end{pmatrix}.$$

10 Suppose first that the limit superior of $s_n^2(r_n/s_n - \kappa)$ is ∞. Since $s_n^2(r_n/s_n - \kappa) < s_n/q_n$ for all n, we have

$$\vartheta(\kappa) = \limsup_{n\to\infty} \frac{s_n}{q_n} \geq \limsup_{n\to\infty} s_n^2\left(\frac{r_n}{s_n} - \kappa\right) = \infty.$$

Suppose conversely that $\vartheta(\kappa) = \infty$. If κ is rational, then there exists a positive integer N with $\phi^{N-1}(\kappa) = 0$; so, $\ell_n(\kappa) = 1$ for all $n \geq N$. This means that the sequence $\{s_n\}$ forms an arithmetic progression with common difference q_N for $n \geq N$. Since $s_n = (n - N)q_N + s_N$ for $n \geq N$, we get

$$\begin{aligned} s_n^2\left(\frac{r_n}{s_n} - \kappa\right) &= s_n^2 \sum_{k=0}^{\infty} \frac{1}{s_{n+k}s_{n+k+1}} \\ &= \frac{s_n^2}{q_N}\left(\left(\frac{1}{s_n} - \frac{1}{s_{n+1}}\right) + \left(\frac{1}{s_{n+1}} - \frac{1}{s_{n+2}}\right) + \cdots\right) \\ &= \frac{s_n}{q_N} = n - N + \frac{s_N}{q_N} \to \infty \end{aligned}$$

as $n \to \infty$. So we assume that κ is irrational in what follows. Since $\vartheta(\kappa) = \infty$, $\Upsilon(\kappa)$ contains a string of consecutive 1's of arbitrary length by Lemma 5.11. Suppose that $\ell_{M-1} \geq 2$ and $\ell_M = \ell_{M+1} = \cdots = \ell_{M+2m-1} = 1$ for some large integers M and m, where $\ell_n = \ell_n(\kappa)$. Hence $s_M, \ldots, s_{M+2m}$ forms an arithmetic progression with common difference q_M and therefore

$$\begin{aligned} s_{M+m}^2\left(\frac{r_{M+m}}{s_{M+m}} - \kappa\right) &> s_{M+m}^2\left(\frac{r_{M+m}}{s_{M+m}} - \frac{r_{M+2m}}{s_{M+2m}}\right) \\ &= s_{M+m}^2\left(\frac{1}{s_{M+m}s_{M+m+1}} + \frac{1}{s_{M+m+1}s_{M+m+2}} + \cdots + \frac{1}{s_{M+2m-1}s_{M+2m}}\right) \\ &= \frac{s_{M+m}^2}{q_M}\left(\left(\frac{1}{s_{M+m}} - \frac{1}{s_{M+m+1}}\right) + \cdots + \left(\frac{1}{s_{M+2m-1}} - \frac{1}{s_{M+2m}}\right)\right) \\ &= \frac{s_{M+m}}{s_{M+2m}}m = \frac{s_M + mq_M}{s_M + 2mq_M}m > \frac{m}{2}. \end{aligned}$$

Since m can be taken to be arbitrarily large, this completes the proof.

Chapter 6

1 Let $\{a_{-2n}\}_{n\geq 0}$ and $\{a_{-1-2n}\}_{n\geq 0}$ be any strictly monotonically decreasing sequences in $(0, 1)$ converging to $\alpha > 0$ and $\beta > a_0$ respectively. Take two constants $\gamma > \delta$ in (a_0, β) suitably. Then, connecting the infinite points

$$(0, \gamma), \quad (\alpha, \beta), \quad \{(a_{-2n}, a_{1-2n})\}_{n\geq 1} \quad \text{and} \quad (a_0, 1)$$

by segments in this order in xy-plane, we obtain a strictly monotonically increasing continuous function $f(x)$ on $[0, a_0)$. Similarly, we can define a strictly monotonically increasing function $f(x)$ on $[a_0, 1]$ by connecting the infinite points

$$(a_0, 0), \quad (\beta, \alpha), \quad \{(a_{1-2n}, a_{2-2n})\}_{n\geq 1} \quad \text{and} \quad (1, \delta)$$

by segments in this order. Clearly we have $f \in \mathfrak{B}_\infty$ with $c_{-n} = a_{-n}$ for all $n \geq 0$ and α, β are periodic points with period 2. When $n \geq 3$, we have $J_1 = [0, c_{2-2[n/2]})$, $J_n = [c_{-1}, 1)$ and so,

$$L = f(J_n) \cup K_1 \cup f(J_1) = [c, c_{3-2[n/2]}),$$

where we understand $c_1 = 1$ exceptionally. Thus $c_{2-n} \in \operatorname{Int} L$ if and only if n is odd for $n \geq 3$. The number of points $c_0, c_{-1}, c_{-3}, \dots, c_{2-n}$ is $(n+1)/2$. If we take $2n+1$ instead of n, we have $q_n = 2n + 1$ and $p_n = n + 1$. The Farey interval of order $2n$ containing p_n/q_n is thus

$$L_n = \left[\frac{1}{2}, \frac{n}{2n-1}\right);$$

so, $\xi_n = n/(2n-1)$. Since $\epsilon_k(0) = 0$ or 1 according as k is even or odd, we have

$$\Theta_f(0, t) = t + t^3 + t^5 + \cdots = \frac{t}{1-t^2}.$$

[2] Let $\{b_{-2n}\}_{n\geq 0}$ and $\{b_{-1-2n}\}_{n\geq 0}$ be any strictly monotonically increasing sequences in $(0, 1)$ converging to $\beta < 1$ and $\alpha < b_0$ respectively. Take two constants $\gamma > \delta$ in (α, b_0) suitably. Then, connecting the infinite points

$$(0, \gamma), \quad \{(b_{1-2n}, b_{2-2n})\}_{n\geq 1}, \quad (\alpha, \beta) \quad \text{and} \quad (b_0, 1)$$

by segments in this order in xy-plane, we obtain a strictly monotonically increasing continuous function $f(x)$ on $[0, b_0)$. Similarly, we can define a strictly monotonically increasing function $f(x)$ on $[b_0, 1]$ by connecting the infinite points

$$(b_0, 0), \quad \{(b_{-2n}, a_{1-2n})\}_{n\geq 1}, \quad (\beta, \alpha) \quad \text{and} \quad (1, \delta)$$

by segments in this order. Clearly we have $f \in \mathfrak{B}_\infty$ with $c_{-n} = b_{-n}$ for all $n \geq 0$ and α, β are periodic points with period 2. When $n \geq 3$, we have $J_1 = [0, c_{-1})$, $J_n = [c_{2-2[n/2]}, 1)$ and so,

$$L = f(J_n) \cup K_1 \cup f(J_1) = [c_{3-2[n/2]}, c),$$

where we understand $c_1 = 0$ exceptionally. Thus $c_{2-n} \in \operatorname{Int} L$ if and only if n is odd for $n \geq 3$. The number of points $c_0, c_{-2}, c_{-4}, \dots, c_{3-n}$ is $(n-1)/2$. If we take $2n+1$ instead of n, we have $q_n = 2n+1$ and $p_n = n$. The Farey interval of order $2n$ containing p_n/q_n is thus

$$L_n = \left[\frac{n-1}{2n-1}, \frac{1}{2}\right);$$

so, $\xi_n = 1/2$. Since $\epsilon_k \circ f(0) = 0$ or 1 according as k is even or odd, we have

$$\Theta_f(0, t) = t^2 + t^4 + t^6 + \cdots = \frac{t^2}{1-t^2}.$$

[3] Use $\lceil x \rceil = [x]$ if $x \in \mathbb{Z}$ and $\lceil x \rceil = [x] + 1$ otherwise.

4 Suppose, on the contrary, that there exists $f \in \mathfrak{B}$ satisfying $\epsilon_n(0) = t_n$ for all $n \geq 0$. Substituting $x = 1$ into

$$\sum_{k=0}^{2^n-1} (-1)^{\sigma_k} x^k = \prod_{k=0}^{n-1} (1 - x^{2^k}),$$

we get

$$\sum_{k=0}^{2^n-1} (-1)^{\sigma_k} = 0; \quad \text{or, equivalently} \quad \sum_{k=0}^{2^n-1} t_k = 2^{n-1}.$$

This implies that $\rho(f) = 1/2$. However it follows from Theorem 4.2 and Lemma 6.3 that the sequence $\{\epsilon_n(0)\}$ does not contain any consecutive 1's.

Chapter 7

1 For each $m \geq 2$ there are infinitely many points, expressed as two different expansions in base m. Let E_k be the set of points x in the interval $[0, 1)$ such that the k-th digit in *every* triadic expansion of x is "1". It is easily seen that $E_1 = (1/3, 2/3)$, which is the first middle third used to define Cantor's ternary set. By the self-similar structure of triadic expansions we find that

$$E_k = \bigcup_{i=1}^{3^{k-1}} \left(\frac{3i-2}{3^k}, \frac{3i-1}{3^k} \right).$$

Therefore

$$[0, 1) \setminus \bigcup_{k=1}^{\infty} E_k,$$

is the set of points whose triadic expansion does not contain the digit "1". We thus obtain Cantor's ternary set by adding the endpoint $x = 1$ to the above remainder set.

2 Suppose that $x \in \Omega_f$ and $f(x) \notin \Omega_f$. Since $\mathrm{Orb}_f(0) \cup \mathrm{Orb}_f(1) \subset \Omega_f$, both $x \neq f^i(0)$ and $x \neq f^i(1)$ hold for all $i \geq 0$. By the assumption, we have $x \in \overline{f^n(J)}$ for all $n \geq 1$. It follows from (iii) of Lemma 4.3 that

$$f^n(J) = J \setminus \bigcup_{k=1}^{n} K_k \quad \text{and} \quad \overline{f^n(J)} = [0, 1] \setminus \bigcup_{k=1}^{n} \mathrm{Int}\, K_k$$

where $J = [0, 1)$; hence $x \in f^n(J)$. Therefore $f(x) \in f^{n+1}(J) \subset \overline{f^{n+1}(J)}$; so, $f(x) \in \Omega_f$, a contradiction.

3 If $f \in \mathfrak{B}_q$, then the proof of Theorem 4.2 implies that every orbit has the same periodic pattern; so, $\eta(0) \sim \eta(1)$. If $f \in \mathfrak{B}_\infty$, then Lemma 4.3 implies that $\eta \circ f(0) = \eta \circ f(1)$.

4 By Lemma 4.3 we see that

$$\bigcap_{n=0}^{\infty} f^n([0,1)) = [0,1)\setminus\bigcup_{n=1}^{\infty} K_n = \Omega_f\setminus \mathrm{Orb}_f(1)$$

is obviously uncountable, which we denote by X. For any $x \in X \subset \Omega_f$ it follows from Exercise 2 in this chapter that $f(x) \in \Omega_f$. If $f(x) = f^k(1)$ for some integer $k \geq 1$, then we have $x = f^{k-1}(1)$, a contradiction; thus, $f(x) \in X$ and $f(X) \subset X$. Conversely, for any $x \in X$ there exists a unique point $y \in [0,1)$ satisfying $x = f(y)$, because $x \in f([0,1))$. If $y \in K_s$ for some $s \geq 1$, then $x = f(y) \in K_{s+1}$, a contradiction. Hence $y \in X$ and $x = f(y) \in f(X)$; so, $X \subset f(X)$. Therefore $f(X) = X$. The maximality is straightforward.

5 For any $0 \leq y < 1$ it follows from (3.2) and (3.3) that

$$\left| \frac{F^n(0)}{n} - \frac{1}{n}\sum_{k=0}^{n-1} \epsilon \circ f^k(y) \right| \leq \frac{2}{n},$$

where F is the lift defined in (3.1). For any $x \in X$, substituting $f^{1-n}(x)$ for y, we have

$$\left| \frac{F^n(0)}{n} - \frac{1}{n}\sum_{i=0}^{n-1} \epsilon \circ f^{-i}(x) \right| \leq \frac{2}{n}.$$

Chapter 8

1 Use the fact that $\lceil kp/q \rceil - 1 = [kp/q]$ for $1 \leq k < q$ and $\lceil kp/q \rceil = [kp/q]$ for $k = q$.

2 We consider the rectangular set of lattice points $\{(i,j) : 0 \leq i \leq q, 0 \leq j \leq p\}$. Note that there are just two points $(0,0)$ and (q,p) on the straight line $px - qy = 0$. So the sum of $w^i z^j$ over lattice points (i,j) on and below the line is

$$\begin{aligned}
\sum_{i=0}^{q} w^i \sum_{j=0}^{[ip/q]} z^j &= \frac{1}{1-z}\sum_{i=0}^{q} w^i (1 - z^{[ip/q]+1}) \\
&= \frac{1 - z + wz - w^{q+1}}{(1-w)(1-z)} - \frac{wz}{1-z} P_{p,q}(w,z).
\end{aligned}$$

Similarly the sum of $w^i z^j$ on and over the line is

$$\begin{aligned}
\sum_{j=0}^{p} z^j \sum_{i=0}^{[jq/p]} w^i &= \frac{1}{1-w}\sum_{j=0}^{p} z^j (1 - w^{[jq/p]+1}) \\
&= \frac{z(1-z^p)}{(1-w)(1-z)} + 1 + w^q z^p - \frac{wz}{1-w} P^+_{q,p}(z,w).
\end{aligned}$$

So, the sum of the above two partial sums is equal to

$$\frac{1-w^{q+1}}{1-w} \cdot \frac{1-z^{p+1}}{1-z} + 1 + w^q z^p,$$

which conduces to the desired identity.

3 We consider the set of lattice points $E = \{(i, j) : i, j \in \mathbb{N}\}$. Define E^- and E^+ as the subsets of E satisfying $i < \mu j$ and $i > \mu j$ respectively. Then the sum of $w^i z^j$ over the E^- is clearly $M_\mu(w, z)$, while the sum over E^+ is

$$\sum_{(i,j)\in E^+} w^i z^j = \sum_{i,j\geq 1, j<i/\mu} w^i z^j = M_{1/\mu}(z, w).$$

The relation between $Q_{1/\mu}(w, z)$ and $Q_\mu(z, w)$ follows from (8.5).

4 The equality follows immediately from (8.3) and (8.4). Substituting $z = 1/m$ into the equality we get

$$(m-1)H_\mu\left(\frac{1}{m}\right) = \sum_{n=1}^{\infty} \frac{1}{m^{[n/\mu]}},$$

giving an expansion in base m of the number $(m-1)H_\mu(1/m)$. Use the fact that the sequence of digits in the expansion in base m of x is eventually periodic if and only if $x \in \mathbb{Q}$.

5 Since $\mu = [\mu] + \{\mu\}$, we have

$$Q_\mu(w, z) = z^{[\mu]} \sum_{k=1}^{\infty} (wz^{[\mu]})^{k-1} z^{[\{\mu\}k]}.$$

6 Since $\varphi^2 - \varphi - 1 = 0$, it follows from the formulae of Exercises 3 and 5 in this chapter that

$$Q_\varphi(w, z) = zQ_{1/\varphi}(wz, z) = \frac{z}{1-wz}(1 - (1-z)Q_\varphi(z, wz)).$$

7 It is not hard to see the following identity by induction on n:

$$Q_\varphi(w, z) = \frac{z}{1-wz} + (1-z)\sum_{k=1}^{n-1}(-1)^k \frac{w^{F_{k+2}-1}z^{F_{k+3}-1}}{(1-w^{F_k}z^{F_{k+1}})(1-w^{F_{k+1}}z^{F_{k+2}})}$$
$$+ (-1)^n(1-z)\frac{w^{F_{n+1}-1}z^{F_{n+2}-1}}{1-w^{F_n}z^{F_{n+1}}} Q_\varphi(w^{F_{n-1}}z^{F_n}, w^{F_n}z^{F_{n+1}}).$$

Since $\left|w^{F_n}z^{F_{n+1}}\right| < |z|^{F_{n+1}-F_n} \to 0$ as $n \to \infty$, the last term in the right-hand side of the above formula tends to 0 as $n \to \infty$.

8 Substitute $w = 1$ to the formula in Exercise 7 in this chapter. Note that $z/(1-z)$ becomes the first term in the first sum.

9 When $a = b$, Theorem 8.7 implies that

$$c = 1 - \frac{1-a}{a} Q_{1/\rho}(1, a),$$

where $\rho = \rho(\psi_{a,a,c})$. Use then the formula

$$M_\rho(1,a) = H_\rho(a) = \frac{a}{1-a} Q_{1/\rho}(1,a).$$

[10] When $a = 1$, Theorem 8.6 implies that

$$c = 1 - \frac{1}{1-b} \cdot \frac{1}{Q_\rho(1,a)} = 1 - \frac{b}{(1-b)^2} \cdot \frac{1}{H_{1/\rho}(b)},$$

where $\rho = \rho(\psi_{1,b,c})$.

[11] Let $\mu = \alpha - 1 > 0$. From Exercise [3] in this chapter we have

$$(1-w)\sum_{n=1}^{\infty} w^{n-1} z^{[n/\mu]} + (1-z)\sum_{n=1}^{\infty} z^{n-1} w^{[\mu n]} = 1 \tag{4}$$

for $|w|^\mu |z| < 1$. Thus, putting $w = z$, we get

$$\frac{1-z}{z}\sum_{n=1}^{\infty} z^{n+[n/\mu]} + \frac{1-z}{z}\sum_{n=1}^{\infty} z^{n+[\mu n]} = 1$$

for $|z| < 1$. This completes the proof, because $\alpha = 1 + \mu$ and $\beta = 1 + 1/\mu$.

[12] Differentiating termwise the identity (4) with respect to w for $k \geq 1$ times and substituting $w = 1$, it can be easily seen that

$$\sum_{n=1}^{\infty} \binom{[\mu n]}{k} z^{n-1} = \frac{z}{1-z} \sum_{n=1}^{\infty} \binom{n-1}{k-1} z^{[n/\mu]} \tag{5}$$

for $|z| < 1$. Since $\binom{n-1}{k-1} = \binom{n}{k} - \binom{n-1}{k}$ and since the identity

$$x^\ell = \sum_{k=1}^{\ell} a_k \binom{x}{k}$$

holds for some coefficients $a_1, \ldots, a_\ell$, it follows from the formula (5) that

$$\sum_{n=1}^{\infty} [\mu n]^\ell z^n = \frac{z}{1-z} \sum_{n=1}^{\infty} (n^\ell - (n-1)^\ell) z^{[n/\mu]}$$

for any $\ell \geq 1$, which completes the proof. We obtain the formula in Exercise [4] in this chapter by putting $P(x) = x$.

Chapter 9

[1] By (iii) of Lemma 9.2, $u_n(r_{a,b})$ is periodic if and only if there exists a positive integer m such that $[\{\kappa n\} + \kappa m]$ is constant for $n \geq 0$. This never happens for any irrational κ. However, if $\kappa = p/q$, then $m = q$ satisfies that property.

[2] Put $\rho = \rho(\psi_{a,b,c})$ and $\kappa = \kappa_{a,b}$. By Lemma 8.1 one has

$$a^{1-\rho} b^{\rho} < 1 = a^{1-\kappa} b^{\kappa},$$

which obviously implies $\rho > \kappa$.

[3] For any

$$a_0 \in \left(a, b + \frac{1-b}{c}\right)$$

put $\psi = \psi_{a,b,c}$ and $\psi_0 = \psi_{a_0,b,c}$. Let F and F_0 be the lifts defined in (3.1) for ψ and ψ_0 respectively. Since $\psi_0(x) \le \psi(x)$ for $0 \le x < 1$, we have $F_0(x) \le F(x)$ for $x \ge 0$; hence ${F_0}^n(x) \le F^n(x)$ for $n \ge 1$ and so $\rho(\psi_0) \le \rho(\psi)$.

[4] For any

$$b_1 \in \left(b, \frac{1-ac}{1-c}\right)$$

put $\psi = \psi_{a,b,c}$ and $\psi_1 = \psi_{a,b_1,c}$. Let F and F_1 be the lifts defined in (3.1) for ψ and ψ_1 respectively. Since $\psi(x) \le \psi_1(x)$ for $0 \le x < 1$, we have $F(x) \le F_1(x)$ for $x \ge 0$; hence $F^n(x) \le {F_1}^n(x)$ for $n \ge 1$ and so $\rho(\psi) \le \rho(\psi_1)$.

[5] For any $b_0 \in (0, b)$ let F and F_0 be the lifts defined in (3.1) for ϕ and $\psi_{a,b_0,c}$ respectively. Since $F(x) \ge F_0(x)$, we have $\rho(\phi) \ge \rho(\psi_{a,b_0,c})$ and hence $\rho(\phi) > \kappa_{a,b_0}$ by the inequality in Exercise [2] in this chapter. Letting $b_0 \to b-$, one has $\rho(\phi) \ge \kappa_{a,b}$.

On the other hand, for any

$$c \in \left(0, \frac{1-b}{a-b}\right)$$

we have $F(x) < F_1(x)$, where F_1 is the lift defined in (3.1) for $\psi_{a,b,c}$. Therefore we get

$$\rho(\phi) \le \rho(\psi_{a,b,c}) = r_{a,b}(c).$$

Since $r_{a,b}(c)$ tends to $\kappa_{a,b}$ as $c \to (1-b)/(a-b)-$, we have $\rho(\phi) \le \kappa_{a,b}$, as required.

[6] For any $n \ge 0$ and $x \in [0, 1)$ we put $x_n = \psi^n_{a,a,c}(x)$ and $\epsilon_n = \mathrm{H}(\psi^n_{a,a,c}(x) - c)$. Since

$$x_{n+1} = 1 - \epsilon_n + a(x_n - c),$$

we have

$$\frac{1-a}{n}\sum_{i=0}^{n-1} x_i + \frac{1}{n}\sum_{i=0}^{n-1} \epsilon_i = 1 - ac + \frac{x_0 + x_n}{n},$$

which obviously implies (i) and (ii). On the other hand, note that

$$\min(1 - ac, c) \le x + \psi_{a,a,c}(x) \le \max(1 + c, 1 + a(1 - c))$$

for any $0 \le x < 1$. Hence we have

$$\frac{1}{2}\min(1 - ac, c) \le \rho^* \le \frac{1}{2}\max(1 + c, 1 + a(1 - c)),$$

which implies (iii) from (ii).

Chapter 10

[1] We define the subset E of $\mathbb{N}$ as the set of integers $[kq/p]$ as $k \in [1, p]$ varies. Similarly define the subset E' by the set of integers $\lceil kq/(q-p) \rceil - 1$ as $k \in [1, q-p]$ varies. Since $q/p > 1$ and $q/(q-p) > 1$, it is obvious that $\#E = p$ and $\#E' = q - p$; hence $\#E + \#E' = q$. Suppose now that there exists some $i \in E \cap E'$. Then

$$\frac{p}{q} i \le k < \frac{p}{q}(i+1) \quad \text{and} \quad \left(1 - \frac{p}{q}\right) i < k' \le \left(1 - \frac{p}{q}\right)(i+1)$$

for some positive integers $k \le p$ and $k' \le q - p$. Adding these inequalities we easily get $i < k + k' < i + 1$, a contradiction. Hence $E \cap E' = \emptyset$ and $E \cup E' = \{1, 2, ..., q\}$; so,

$$P_{q,p}(1,z) + P^{+}_{q,q-p}(1,z) = \sum_{k \in E} z^k + \sum_{k \in E'} z^k = z + z^2 + \cdots + z^q.$$

[2] By definition we have

$$\begin{aligned} P_{kp,kq}(w,z) &= \sum_{\ell=1}^{kq} w^{\ell-1} z^{[\ell p/q]} = \sum_{s=0}^{k-1} w^{sq} z^{sp} \sum_{r=1}^{q} w^{r-1} z^{[rp/q]} \\ &= \frac{1 - w^{kq} z^{kp}}{1 - w^q z^p} P_{p,q}(w,z). \end{aligned}$$

We get the similar result for $P^{+}_{kp,kq}(w,z)$.

[3] Using $-[x] = \lceil -x \rceil$ we have

$$w^{q-1} z^p P_{p,q}\left(\frac{1}{w}, \frac{1}{z}\right) = \sum_{k=1}^{q} w^{q-k} z^{p + \lceil -kp/q \rceil} = 1 - w^q z^p + wz P^{+}_{p,q}(w,z).$$

[4] By termwise integration it follows from (10.14) that

$$\sum_{n=2}^{\infty} a_n \varphi(n) = 1 \quad \text{where} \quad a_n = \int_0^1 \frac{x^{n-2}(1-x)^2}{1-x^n} dx.$$

Again, by termwise integration, we obtain

$$a_n = \sum_{k=1}^{\infty} \int_0^1 x^{kn-2}(1-x)^2 dx = \sum_{k=1}^{\infty} \left(\frac{1}{kn-1} - \frac{2}{kn} + \frac{1}{kn+1} \right).$$

Therefore, applying the formula (10.15) we get

$$na_n = -\Psi\left(1 + \frac{1}{n}\right) - \Psi\left(1 - \frac{1}{n}\right) - 2\mathrm{C}.$$

For examples, $a_2 = 2\log 2 - 1$, $a_3 = \log 3 - 1$, $a_4 = \dfrac{3}{2}\log 2 - 1$ and

$$a_5 = \frac{\sqrt{5}}{10}\log\frac{5+\sqrt{5}}{5-\sqrt{5}} - 1.$$

[5] The set E endowed with uniform norm $\|\cdot\|$ becomes a complete metric space. By definition it is easily seen that

$$\|T(f) - T(g)\| \le \frac{1}{2}\|f - g\|;$$

so, $T : E \to E$ is a contraction. Therefore there exists a unique fixed point of T. (See Exercise [2] in Chapter 1.) Cantor's function is obviously the fixed point of T.

Chapter 11

[1] Using $P_{p,q}(1,1) = P^+_{p,q}(1,1) = q$ we have $\alpha_{p/q}(1,1) = \beta_{p/q}(1,1) = 1 - p/q$. We next have

$$\lim_{z\to\infty}\alpha_{p/q}(1,z) = 1 - \lim_{z\to\infty}\frac{\sum_{k=1}^{p} z^{k-1}}{\sum_{k=1}^{q} z^{\lceil kp/q\rceil - 1}} = 1 - \frac{1}{d},$$

where d is the number of positive integers $k \le q$ satisfying $\lceil kp/q\rceil = p$; that is,

$$p - 1 < \frac{p}{q}k \le p.$$

Then it is easily seen that $d = \lceil q/p\rceil$. The last formula is straightforward, because

$$\beta_{p/q}(1,z) = 1 - \frac{\sum_{k=1}^{p} z^{k-1}}{\sum_{k=1}^{q} z^{\lfloor kp/q\rfloor}} = 1 + O\left(\frac{1}{z}\right)$$

as $z \to \infty$.

[2] We have $\alpha_{p/q}(1,z) = 1 - g(z)/f(z)$, where

$$f(z) = \sum_{k=1}^{q} z^{\lceil kp/q\rceil - 1} \quad \text{and} \quad g(z) = 1 + z + \cdots + z^{p-1}.$$

Using $f(1) = q$, $g(1) = p$, $g'(1) = p(p-1)/2$ and

$$f'(1) = \sum_{k=1}^{q}\left(\left\lceil \frac{p}{q}k\right\rceil - 1\right) = \sum_{k=1}^{q}\left\lfloor \frac{p}{q}k\right\rfloor - 1 = \frac{(p-1)(q+1)}{2},$$

we easily get

$$\lim_{z\to 1}\frac{d}{dz}\alpha_{p,q}(1,z) = \frac{p(p-1)}{2q^2}. \tag{6}$$

Now putting

$$f'(z)g(z) - f(z)g'(z) = \sum_{k=0}^{2p-3} a_k z^k,$$

we have

$$a_k = \sum_{\lceil ip/q \rceil + j - 3 = k} \left(\left\lceil \frac{p}{q} i \right\rceil - j \right)$$

where the sum in the right-hand side extends over all pair (i, j) satisfying $1 \le i \le q$, $1 \le j \le p$ and $\lceil ip/q \rceil + j - 3 = k$; so a_k can be rewritten as

$$a_k = \sum_{\substack{1 \le i \le q \\ k+3-p \le \lceil ip/q \rceil \le k+2}} \left(2\left\lceil \frac{p}{q} i \right\rceil - k - 3 \right). \tag{7}$$

Suppose that there exist integers i and k satisfying $2\lceil ip/q \rceil < k+3$. Since $k \le 2p-3$, we must have $i < q$. Put

$$i^* = \left\lceil \frac{q}{p}(k+2) \right\rceil - i.$$

We then show the following three properties:

(i) $i < i^* \le q$.

Since $\lceil ip/q \rceil < (k+3)/2$ and $(k+3)/2$ is either an integer or a half-integer, we have $\lceil ip/q \rceil \le k/2 + 1$; so,

$$\left\lceil \frac{q}{p}(k+2) \right\rceil \ge \frac{q}{p}(k+2) \ge \frac{2q}{p}\left\lceil \frac{p}{q} i \right\rceil > 2i,$$

because $ip/q \notin \mathbb{Z}$. This means that $i^* > i$. On the other hand, since $k + 3 - p < ip/q + 1$, we have $k + 2 < ip/q + p$; hence, $(k+2)q/p < q + i$. This implies that $\lceil (k+2)q/p \rceil \le q + i$, as required.

(ii) $\lceil ip/q \rceil + \lceil i^* p/q \rceil$ is either $k+3$ or $k+4$.

Since $i + i^* = \lceil (k+2)q/p \rceil$, we obtain

$$k + 2 \le \frac{p}{q}(i + i^*) < k + 2 + \frac{p}{q}.$$

Putting $ip = sq + r$, $i^* p = s^* q + r^*$ with $1 \le r < q$, $0 \le r^* < q$, we have

$$N \le \frac{r + r^*}{q} < N + \frac{p}{q},$$

where $N = k + 2 - s - s^*$. Therefore N is either 0 or 1. Then

$$\left\lceil \frac{p}{q} i \right\rceil + \left\lceil \frac{p}{q} i^* \right\rceil = k + 3 + \left\lceil \frac{r^*}{q} \right\rceil - N.$$

However $r^* = 0$ (that is, $i^* = q$) and $N = 1$ do not occur simultaneously.

(iii) $k+3-p \le \lceil i^*p/q \rceil \le k+2$.

Obviously the first inequality follows from the property (i). If $i^* = q$, then

$$q+1 \le i+i^* = \left\lceil \frac{q}{p}(k+2) \right\rceil < \frac{q}{p}(k+2)+1;$$

so, we have $p < k+2$. On the other hand, if $i^* < q$, then $\lceil i^*p/q \rceil = s^*+1$. Suppose, on the contrary, that $s^* \ge k+2$. Since $N \ge 0$, we get $s^* = k+2$ and therefore

$$i+i^* = \left\lceil \frac{q}{p}s^* \right\rceil = \left\lceil i^* - \frac{r^*}{p} \right\rceil \le i^*,$$

because $r^* \ge 1$, a contradiction. Thus we have $\lceil i^*p/q \rceil \le k+2$.

The properties (i), (ii) and (iii) imply that, if a negative term appears in the sum in the formula (7) for some i,k, then there exists a positive term corresponding to i^*, k satisfying

$$\left(2\left\lceil \frac{p}{q}i \right\rceil - k - 3\right) + \left(2\left\lceil \frac{p}{q}i^* \right\rceil - k - 3\right) \ge 0.$$

Let $\mathfrak{I} = \{i_1, i_2, \ldots, i_m\}$ be the set of i's corresponding to negative terms in the sum in the formula (7) and let $\mathfrak{I}^* = \{i_1^*, i_2^*, \ldots, i_m^*\}$. We have

$$\#\mathfrak{I}^* = m \quad \text{and} \quad \mathfrak{I} \cap \mathfrak{I}^* = \emptyset,$$

because $\mathfrak{I}^*$ is the subset corresponding to positive terms. Since $2\lceil ip/q \rceil - k - 3$ is monotonically increasing with respect to i, it follows that $\mathfrak{I} < \mathfrak{I}^*$. We thus conclude that all the coefficients a_k are non-negative and the function $f(z)/g(z)$ is monotonically increasing for $z \ge 0$. If $a_k = 0$ for all k, then the function $\alpha_{p/q}(1,z)$ becomes a constant, contrary to the formula (6) when $p \ge 2$. Hence $\alpha_{p/q}(1,z)$ is strictly monotonically increasing for $z \ge 0$.

3 We have $\beta_{p/q}(1,z) = 1 - g_0(z)/f_0(z)$, where

$$f_0(z) = \sum_{k=1}^{q} z^{\lfloor kp/q \rfloor} \quad \text{and} \quad g_0(z) = 1 + z + \cdots + z^{p-1}.$$

Using $f_0(1) = q$ and

$$f_0'(1) = \sum_{k=1}^{q} \left\lfloor \frac{p}{q}k \right\rfloor = \frac{p(q+1)-q+1}{2},$$

we get

$$\lim_{z \to 1} \frac{d}{dz}\beta_{p,q}(1,z) = \frac{p(p+1)}{2q^2}. \tag{8}$$

Just like the proof of Exercise 2 in this chapter, let b_k be the coefficient of z^k of the polynomial $f_0'(z)g_0(z) - f_0(z)g_0'(z)$; that is,

$$b_k = \sum_{\lfloor ip/q \rfloor + j - 2 = k} \left(\left\lfloor \frac{p}{q}i \right\rfloor - j + 1 \right)$$

for $0 \le k \le 2p-2$, where the sum in the right-hand side extends over all pair (i, j) satisfying $1 \le i \le q$, $1 \le j \le p$ and $[ip/q] + j - 2 = k$; so b_k can be rewritten as

$$b_k = \sum_{\substack{1 \le i \le q \\ k+2-p \le [ip/q] \le k+1}} \left(2\left[\frac{p}{q}i\right] - k - 1\right).$$

Suppose now that there exist integers i and k satisfying $2[ip/q] < k+1$. Since $k \le 2p-2$, we must have $i < q$. Put

$$i^{**} = \left[\frac{q}{p}(k+2)\right] - i + 1.$$

We then show the following three properties:

(i) $i < i^{**} \le q$.

Since $ip/q - 1 < [ip/q] \le k/2$, we have

$$\left[\frac{q}{p}(k+2)\right] > \frac{q}{p}(k+2) - 1 > 2i - 1.$$

This means that $i^{**} > i$. On the other hand, since $k + 2 - p < ip/q$, we obtain $(k+2)q/p < q + i$; hence, $[(k+2)q/p] \le q + i - 1$. Hence $i^{**} \le q$.

(ii) $[ip/q] + [i^{**}p/q]$ is either $k+1$ or $k+2$.

Since $i + i^{**} = [(k+2)q/p] + 1$, we have

$$\frac{q}{p}(k+2) < i + i^{**} \le \frac{q}{p}(k+2) + 1.$$

Putting $ip = sq + r$, $i^{**}p = s^{**}q + r^{**}$ with $1 \le r < q$, $0 \le r^{**} < q$, we have

$$N_0 \le \frac{r + r^{**}}{q} < N_0 + \frac{p}{q},$$

where $N_0 = k + 2 - s - s^{**}$. Therefore N_0 is either 0 or 1. Then

$$\left[\frac{p}{q}i\right] + \left[\frac{p}{q}i^{**}\right] = s + s^{**} = k + 2 - N_0.$$

(iii) $k + 2 - p \le [i^{**}p/q] \le k + 1$.

The first inequality follows from the property (i) above. Suppose next that

$$\left[\frac{p}{q}i^{**}\right] = k + 2 - N_0 - \left[\frac{p}{q}i\right] \ge k + 2.$$

Then we have $N_0 = [ip/q] = 0$ and so, $s = 0$, $r = ip \ge p$. On the other hand, $r \le r + r^{**} < p$, a contradiction. Hence $[i^{**}p/q] \le k + 1$, as required.

Therefore, by the similar argument as in the previous proof, we see that all the coefficients b_k is non-negative; thus the function $f(z)/g(z)$ is monotonically increasing for $z \ge 0$. If $b_k = 0$ for all k, then the function $\beta_{p/q}(1, z)$ becomes a constant, contrary to the formula (8). Therefore $\beta_{p/q}(1, z)$ is strictly monotonically increasing for $z \ge 0$.

4 Use the formula in Exercise 1 of Chapter 10.

5 If $0 \le w \le 1$, then

$$1 - \alpha_{1/q}(w, z) = \frac{w^{q-1}}{1 + w + \cdots + w^{q-1}}$$

attains its maximum at $w = 1$. On the other hand, if $w > 1$, then clearly

$$A^-_{0/1}(w, z) - \alpha_{1/q}(w, z) = \frac{1}{w(1 + w + \cdots + w^{q-1})} < \frac{1}{q}.$$

6 If $wz \le 1$, then

$$\alpha_{p/(p+1)}(w, z) \le \frac{(wz)^{p-1}}{1 + wz + \cdots + (wz)^{p-1}}$$

and the right-hand side attains its maximum at $wz = 1$. On the other hand, if $wz > 1$, then

$$\alpha_{p/(p+1)}(w, z) - A^+_{1/1}(w, z) = \frac{1}{w(wz + z - 1)(1 + wz + \cdots + (wz)^{p-1} + w^p z^{p-1})},$$

which is less than $1/p$ because $w(wz + z - 1) > wz > 1$.

7 Using Exercises 1 and 2 in Chapter 8 and the identity (10.6) we obtain

$$\begin{aligned} &P^+_{p,q}(w, z)\, P_{s,r}(z, w) - P_{q,p}(z, w) P^+_{r,s}(w, z) \\ &\quad = w^{q+s-1} z^{p+r-1} + w^{q-1} z^{p-1}\Big((1 - z) P_{s,r}(z, w) + (1 - w) P^+_{r,s}(w, z)\Big) \\ &\quad = w^{q-1} z^{p-1}, \end{aligned}$$

which is the identity given in (11.1). Similarly we have

$$\begin{aligned} &P_{p,q}(w, z)\, P^+_{s,r}(z, w) - P^+_{q,p}(z, w) P_{r,s}(w, z) \\ &\quad = w^{q+s-1} z^{p+r-1} + w^{s-1} z^{r-1}\Big((1 - w) P_{p,q}(w, z) + (1 - z) P^+_{q,p}(z, w)\Big) \\ &\quad = w^{s-1} z^{r-1}, \end{aligned}$$

which is the identity given in (11.6).

Chapter 12

1 Use the relation $s_{n+1} = q_{n+1} + s_n$.

2 For any $\varepsilon > 0$ take a sufficiently large integer q_0 satisfying

$$q_0^\varepsilon > 2^m \mu^{1-1/m}, \quad q_0^{m\omega} > \mu^{-1/m} \quad \text{and} \quad \left|\mu - \frac{u}{v}\right| \ge \frac{1}{v^{\omega+\varepsilon}}$$

for all $u \in \mathbb{Z}$ and $v \geq q_0^m$. Then for any $p \in \mathbb{Z}$ and $q \geq q_0$ with $|p/q| \leq 2\mu^{1/m}$ we have

$$\left|\mu^{1/m} - \frac{p}{q}\right| = \left|\frac{\mu - p^m/q^m}{\sum_{i=0}^{m-1} \mu^{i/m}(p/q)^{m-1-i}}\right| \geq \frac{1}{2^m \mu^{1-1/m} q^{m(\omega+\varepsilon)}} > \frac{1}{q^{m\omega+(m+1)\varepsilon}}.$$

This is valid even for any $p \in \mathbb{Z}$ and $q \geq q_0$ with $|p/q| > 2\mu^{1/m}$, because

$$\left|\mu^{1/m} - \frac{p}{q}\right| \geq \left|\frac{p}{q}\right| - \mu^{1/m} \geq \mu^{1/m} > \frac{1}{q_0^{m\omega}} \geq \frac{1}{q^{m\omega}}.$$

Since ε is arbitrary, this means that $\mu^{1/m}$ has an irrationality measure $m\omega$.

3 For any $n \geq 1$ put $N = [(n+1)/2]$ and

$$a_n = \sum_{k=0}^{N-1} \binom{n-k-1}{k};$$

so, we have

$$a_n = \sum_{k=0}^{N-1} \frac{1}{2\pi i} \int_C \frac{(z+1)^{n-k-1}}{z^{k+1}} dz = \frac{1}{2\pi i} \int_C \frac{(z+1)^n}{z^2+z-1}\left(1 - \frac{1}{z^N(z+1)^N}\right) dz,$$

where C is the circle centered at the origin with radius less than $1/\varphi$, $\varphi = (\sqrt{5}+1)/2$. Since $z = 0$ is the regular point of the rational function $(z+1)^n/(z^2+z-1)$, we get

$$a_n = -\frac{1}{2\pi i} \int_C \frac{(z+1)^{n-N}}{z^N(z^2+z-1)} dz.$$

Using the fact that $-\varphi$ and $1/\varphi$ are roots of the quadratic equation $z^2 + z - 1 = 0$ and that

$$(N+2) - (n-N) = 2 + \left[\frac{n+1}{2}\right] - \left[\frac{n}{2}\right] \geq 2,$$

we see by the residue theorem that

$$a_n = \operatorname*{Res}_{z=-\varphi} f(z) + \operatorname*{Res}_{z=1/\varphi} f(z) \quad \text{where} \quad f(z) = \frac{(z+1)^{n-N}}{z^N(z+\varphi)(z-1/\varphi)}.$$

Since

$$\operatorname*{Res}_{z=-\varphi} f(z) = \frac{(1-\varphi)^{n-N}}{(-\varphi)^N(-\varphi-1/\varphi)} \quad \text{and} \quad \operatorname*{Res}_{z=1/\varphi} f(z) = \frac{\varphi^N(1+1/\varphi)^{n-N}}{\varphi+1/\varphi},$$

it follows from Binet's formula that

$$a_n = \frac{1}{\sqrt{5}}\left(\varphi^n - \left(-\frac{1}{\varphi}\right)^n\right) = F_n,$$

as required.

4 Suppose, on the contrary, that $P_{p,q}(w_0, z_0) = P^+_{q,p}(z_0, w_0) = 0$ for some w_0, z_0. Then it follows from the identity of Exercise 3 in Chapter 8 that $w_0^q z_0^p = 1$. We have, in particular, $w_0 \neq 0$ and $z_0 \neq 0$.

On the other hand, $p/q \neq 1/1$, because $P^+_{1,1}(w, z) = 1$. If $p < q$, then $w_0^{s-1} z_0^{r-1} = 0$ for some fraction r/s by (11.6). If $p > q$, then we obtain $w_0^{q-1} z_0^{p-1} = 0$ by exchanging p and q, also w_0 and z_0 in the identity (11.1). This contradiction completes the proof.

5 We assume that $w < -1, z > 0$ and $1 + wz < 0$. Let p/q be an irreducible fraction p/q in the interval $(0, 1)$. If q is even, then we have

$$P^+_{p,q}(w, z) = \sum_{k=1}^{q} w^{k-1} z^{\lceil kp/q \rceil - 1} = \sum_{\ell=0}^{q/2-1} (1 + wz^{m(\ell)}) w^{2\ell} z^{n(\ell)},$$

where

$$m(\ell) = \left\lceil \frac{(2\ell + 2)p}{q} \right\rceil - \left\lceil \frac{(2\ell + 1)p}{q} \right\rceil \quad \text{and} \quad n(\ell) = \left\lceil \frac{(2\ell + 1)p}{q} \right\rceil - 1.$$

Since $m(\ell) = 0$ or 1, we get

$$1 + wz^{m(\ell)} \leq 1 + w \min(1, z) < 0,$$

which implies that $P^+_{p,q}(w, z) < 0$. On the other hand, if q is odd, then

$$P^+_{p,q}(w, z) = 1 + \sum_{\ell=1}^{(q-1)/2} (1 + wz^{m'(\ell)}) w^{2\ell-1} z^{n'(\ell)},$$

where

$$m'(\ell) = \left\lceil \frac{(2\ell + 1)p}{q} \right\rceil - \left\lceil \frac{2\ell p}{q} \right\rceil \quad \text{and} \quad n'(\ell) = \left\lceil \frac{2\ell p}{q} \right\rceil - 1.$$

Since $m'(\ell) = 0$ or 1, we get $P^+_{p,q}(w, z) > 0$. We obtain the similar result for $P_{p,q}(w, z)$.

6 We assume that $|w| < 1/2$ and $|wz| < 1/2$. Then we have

$$\left| P^+_{p,q}(w, z) \right| = \left| \sum_{k=1}^{q} w^{k-1} z^{\lceil kp/q \rceil - 1} \right| \geq 1 - \sum_{k=2}^{q} |w|^{k-1} |z|^{\lceil kp/q \rceil - 1}$$
$$= 1 - \sum_{k=2}^{q} |w|^{k - \lceil kp/q \rceil} |wz|^{\lceil kp/q \rceil - 1} \geq 1 - \sum_{k=2}^{q} \frac{1}{2^{k-1}} > 0.$$

We have the similar result for $P_{p,q}(w, z)$.

Table of $P_{p,q}(w, z)$

The polynomial $P^+_{p,q}(w, z)$ can be obtained from $P_{p,q}(w, z)$ by replacing the last term $w^{q-1}z^p$ by $w^{q-1}z^{p-1}$ when $q \geq 1$. (See Exercise 1 in Chapter 8.)

p, q	$P_{p,q}(w, z)$	$P^+_{p,q}(w, z)$
$0, 1$	1	$\frac{1}{z}$
$1, 0$	$\frac{w-1}{z}$	0
$1, n$	$1 + w + \cdots + w^{n-2} + w^{n-1}z$	
$n, 1$	z^n	
$2, 2n+1$	$(1 + w + \cdots + w^{n-1})(1 + w^n z) + w^{2n}z^2$	
$2n+1, 2$	$z^n(1 + wz^{n+1})$	
$3, 3n+1$	$(1 + w + \cdots + w^{n-1})(1 + w^n z + w^{2n}z^2) + w^{3n}z^3$	
$3n+1, 3$	$z^n(1 + wz^n + w^2 z^{2n+1})$	
$3, 3n+2$	$(1 + w + \cdots + w^{n-1})(1 + w^n z + w^{2n+1}z^2) + w^{2n}z + w^{3n+1}z^3$	
$3n+2, 3$	$z^n(1 + wz^{n+1} + w^2 z^{2n+2})$	
$4, 4n+1$	$(1 + w + \cdots + w^{n-1})(1 + w^n z + w^{2n}z^2 + w^{3n}z^3) + w^{4n}z^4$	
$4n+1, 4$	$z^n(1 + wz^n + w^2 z^{2n} + w^3 z^{3n+1})$	
$4, 4n+3$	$(1 + w + \cdots + w^{n-1})(1 + w^n z + w^{2n+1}z^2 + w^{3n+2}z^3)$ $+ w^{2n}z + w^{3n+1}z^2 + w^{4n+2}z^4$	
$4n+3, 4$	$z^n(1 + wz^{n+1} + w^2 z^{2n+2} + w^3 z^{3n+3})$	

Bibliography

Ahlfors, L.V. (1979). Complex Analysis, Third Edition, McGraw-Hill International Editions.

Anderson, J.A., Pellionisz, A., and Rosenfeld, E. (1990). Neurocomputing 2, Directions for Research, MIT Press.

Besicovitch, A.S. and Taylor, S.J. (1954). On the complementary intervals of a linear closed set of zero Lebesgue measure, *J. London Math. Soc.* **29**, pp.449 – 459.

Caianiello, E.R. (1961). Outline of a theory of thought process and thinking machines, *J. Theoretical Biology* **1**, pp.204 – 235.

Devaney, R.L. (1989). An Introduction to Chaotic Dynamical Systems, Second Edition, Addison-Wesley.

Falconer, K.J. (1985). The Geometry of Fractal Sets, Cambridge Univ. Press.

Frame, M., Johnson, B. and Sauerberg, J. (2000). Fixed points and Fermat: A dynamical systems approach to number theory, *Amer. Math. Monthly* **107**, no. 5, pp.422 – 428.

Hardy, G.H. and Wright, E.M. (1979). An Introduction to the Theory of Numbers, Fifth Edition, Oxford University Press.

Hata, M. (1982). Dynamics of Caianiello's equation, *J. Math. Kyoto Univ.* **22**, pp.155 – 173.

Hata, M. (1998). Shinkei-kairo-moderu no Kaosu (in Japanese), Asakura Shoten.

Hecke, E. (1921). Uber analytische Funktionen und die Verteilung von Zahlen mod. Eins, *Hamb. Abh.* **1**, pp.54 – 76.

Levine, L. (1999). Fermat's little theorem: A proof by function iteration, *Math. Magazine* **72**, no. 4, pp.308 – 309.

Levinson, N (1966). Wiener's life, *Bull. Amer. Math. Soc.* **72**, no. 1, part II, pp.1 – 32.

Mahler, K. (1982). Fifty years as a mathematician, *J. Number Theory* **14**, pp.121 – 155.

Milnor, J. and Thurston, W. (1988). On iterated maps of the interval, Lecture Notes in Math., vol. 1342, pp.465–563, Springer-Verlag.

Nagumo, J. and Sato, S. (1972). On a responce characteristic of a mathematical neuron model, *Kybernetik* **10**, pp.155 – 164.

Nishioka, K. (1996). Mahler Functions and Transcendence, Lecture Notes in Math., vol. 1631, Springer-Verlag.

Niven, I. and Zuckerman, H.S. (1960). An Introduction to the Theory of Numbers, John Wiley & Sons.

Roth, K. F. (1955). Rational approximations to algebraic numbers, *Mathematika* **2**, pp. 1–20; corrigendum, 168.

Smale, S. (1967). Differentiable dynamical systems, *Bull. Amer. Math. Soc.* **73**, pp. 747–817.

Targonski, G. (1981). Topics in Iteration Theory, Studia Mathematica, Skript 6, Vandenhoeck & Ruprecht.

Index

Printed in the United States
By Bookmasters